Organismic Animal Biology

T0092122

Organismic Animal Biology

An Evolutionary Approach

Ariel D. Chipman

Professor, Department of Ecology, Evolution and Behavior, The Hebrew University of Jerusalem, Israel

Great Clarendon Street, Oxford, OX2 6DP,
United Kingdom

Oxford University Press is a department of the University of Oxford.
It furthers the University's objective of excellence in research, scholarship,
and education by publishing worldwide. Oxford is a registered trade mark of
Oxford University Press in the UK and in certain other countries

© Ariel D. Chipman 2024

The moral rights of the author have been asserted

All rights reserved. No part of this publication may be reproduced, stored in
a retrieval system, or transmitted, in any form or by any means, without the
prior permission in writing of Oxford University Press, or as expressly permitted
by law, by licence or under terms agreed with the appropriate reprographics
rights organization. Enquiries concerning reproduction outside the scope of the
above should be sent to the Rights Department, Oxford University Press, at the
address above

You must not circulate this work in any other form
and you must impose this same condition on any acquirer

Published in the United States of America by Oxford University Press
198 Madison Avenue, New York, NY 10016, United States of America

British Library Cataloguing in Publication Data
Data available

Library of Congress Control Number: 2023944331

ISBN 9780192893581
ISBN 9780192893598 (pbk.)

DOI: 10.1093/oso/9780192893581.001.0001

Printed and bound by
CPI Group (UK) Ltd, Croydon, CR0 4YY

Cover image: Netta Kasher

Links to third party websites are provided by Oxford in good faith and
for information only. Oxford disclaims any responsibility for the materials
contained in any third party website referenced in this work.

Contents

About this book

Who is this book for?

Organismic Animal Biology is an introductory text-book for students just beginning their studies in biology. It is aimed at students with minimal biological training, who might have taken an introductory cell biology course previously, or are taking one in parallel, but who have not studied zoology or evolution. The book is designed to be used as a companion to a course such as Biology 101 or Introduction to Zoology. The book gives a broad background in organismic animal biology that should be part of the basic training of any biologist. It can serve as a first sampling of animal diversity for students intending to specialize in zoology and related disciplines. However, the main target audience is those students who will continue in other fields such as genetics, bio-medical sciences, or brain science. This book will provide them with an understanding of whole organisms, how they function, and why they are built the way they are. This understanding will provide a context for the biology underlying any question the students ultimately choose to study, be it in industry, academia, or policy.

How is this book different from other zoology textbooks?

This book breaks away from the standard structure that most introductory courses have been following almost unchanged for decades. Rather than providing a long list of animal groups with numerous details about their anatomy or physiology, this book focuses on principles. The aim of the book is to provide students with tools to understand animal diversity rather than require them to memorize details. The underlying theme of the book is explicitly evolutionary. By explaining the evolutionary processes that led to the features that animals display, we hope the student will find it easier to follow the principles and put them in context. When the students come across an unfamiliar organism in the future, they will be able to recognize the general aspects of its structure and function, while understanding its specific peculiarities.

About the book's structure

There are two types of chapters in this book, which can be called "principles" chapters and "zoology" chapters. The first few chapters present introductory organismic and evolutionary principles. From there on the two chapter types mostly alternate. The "principles" chapters present concepts or ideas, or else focus on a specific organismic function or organ system. The principles are demonstrated with examples. In the early parts of the book, before the reader has been introduced to different groups of animals, the examples are mostly given from mammals and other vertebrates, which the reader has probably encountered in the non-academic world. As the book progresses and the reader becomes familiar with more and more of the diversity of life, examples come from the groups already introduced.

The "zoological" chapters each introduce a specific group of animals, usually at the level of a phylum. The group in question is introduced in a more-or-less uniform format, starting with the general body plan, moving on to specific organ systems, with an emphasis on the principles already covered, and ending with an overview of the diversity within the group and the evolutionary history of the group.

In addition to the main chapters, some chapters include a box or two, which briefly introduce a topic that is related to the topic of the chapter, but do not fit in the chapter directly. These topics are either short discussions of important concepts, brief introductions to animal groups that for various reasons do not get a full chapter devoted to them, or overviews of crucial events in the history of life on Earth.

The book includes a glossary with close to 600 terms, which appears at the very beginning, even before the first chapter. All terms that appear in the glossary are marked in **bold** upon their first appearance in the text (and sometimes upon their second appearance if they are reintroduced in a different context).

At the front of the book, the reader will find a phylogenetic tree of all the animal groups that are discussed in detail in the book, with their evolutionary relationships. At the very end of the book is a timeline of the geological eras and periods in the history of Earth, some of which are referred to in the text.

About the illustrations

There are close to 100 original illustrations prepared especially for this book by Netta Kasher. The illustrations are intentionally schematic and simplified, in bearing with the main theme of the book, which is to emphasize principles over details. The description of most animal groups is accompanied by an illustration of a lateral view of the main organ systems in the animal and/or a cross section of the internal anatomy. These illustrations are not meant to represent a specific member of the group, or even a specific subgroup, but are a synthesis of features found within it. The anatomical illustrations follow a consistent color coding, with similar organ systems appearing in the same colors in all illustrations, to allow easy comparison among the groups. Other illustrations present a general concept in an "infographic" style. In addition to these illustrations, each one of the groups discussed is represented by full color photos of a sample of their diversity.

Acknowledgments

This book was written over a period of three years, but most of the writing was done in a number of short and intense "writing retreats." During those retreats, I was hosted by colleagues at a number of institutions, and I want to start by thanking those colleagues for accommodating me, providing suitable conditions, and being available for discussions. My hosts over the years were: the Paleobiology Research Group at the School of Earth Sciences at the University of Bristol, the Department of Ecology & Evolution and the Institute of Earth Sciences at the University of Lausanne, the Department of Organismal and Evolutionary Biology and the Museum of Comparative Zoology at Harvard University, and the Department of Integrative Zoology at the University of Vienna.

Each chapter of the book was read and commented upon by two or three additional readers, usually specialists in the topic of the chapter. I want to mention and thank them all. This book would not have been as precise and up to date without their help. In some cases, which I highlight below, their comments helped improve the relevant chapters significantly. Of course, any remaining errors and omissions are entirely my fault.

Wallace Arthur, friend, former collaborator, and experienced book author, provided writing advice and encouragement throughout the writing process and read more than a third of the chapters, especially the conceptual and principles-focused chapters. My close colleague Efrat Gavish-Regev read Chapters 2 and 21 and highlighted many important points. Carl Simpson read Chapters 1 and 10. Thibaut Brunet read Chapters 3 and 5, and his ideas on the evolution of multicellularity feature prominently in the book. Gerhard Scholtz read Chapters 4 and 7, and I also owe him a lot for his role in helping develop my thinking on arthropod evolution. Arnau Sebé-Pedrós read Chapter 5. Sally Leys and Micha Ilan both read Chapter 6 on sponges and improved it considerably. Jeff Camhi, my first scientific mentor, read Chapters 8 and 12. Uli Technau and Mark Martindale both read Chapter 9 on diploblastic organisms, highlighting the open questions and disagreements in the field. Uli also read Chapter 8. Yossi Heller, my first zoology teacher and malacologist extraordinaire, read Chapters 11, 15, and 22. My Institute colleague Michael Brandeis read Chapter 12 on sensory systems, while preparing a course on the same subject. Peter Olson read Chapters 13 and 14 on platyhelminths and parasitism and contributed significantly to the improvement of these chapters, with his expertise on parasitic flatworms. Andreas Wanninger read Chapter 15 on molluscs critically and helped focus the evolutionary questions surrounding this fascinating group. He also read Chapter 19. Ken Halanych read Chapter 16. Patricia Álvarez Campos and Eduardo Zattara both read Chapter 17 on annelids and provided a lot of useful data and insights. Eduardo also read Chapter 10, and Patricia also read Chapter 25. Over the years, Giuseppe Fusco and I have had numerous discussions on the evolution of segmental body plans. He was a natural choice as an external reader for Chapter 18. We still disagree on many points, but the chapter tries to accommodate a synthesis of both our views. Giuseppe also read Chapter 23. Greg Edgecombe, also a friend and collaborator and a master of arthropod biology, read Chapters 20 and 23 and parts of Chapter 13. Allison Daley read parts of Chapter 19 and parts of Chapter 20. Prashant Sharma Read and commented extensively on Chapters 21 and 24. Rich Mooi read

Chapters 24 and 26. His expertise on echinoderm evolution significantly improved that chapter. Jeffrey Thompson also read Chapter 26 and his expertise complemented that of Rich. Noa Shenkar read Chapter 27. Billie Swalla read Chapters 27 and 32 and provided support and encouragement throughout the writing process (and in many other cases in the past). Andreas Hejnol read Chapter 28. Ram Reshef Read Chapters 28, 29, and 31. Mike Benton read Chapters 29 and 30. Anjali Goswami, together with her former and current lab members, including Emily Watt and Agnese Lanzetti, read Chapters 29 and 30. Last but not least, Dale Frank read Chapter 31.

A draft version of the book was tested on the students of my first-year undergraduate course "An Introduction to Organismic Biology" at the Hebrew University of Jerusalem, in the 2022–2023 school year. I have been teaching this course for 15 years, and the book is largely based on my experience teaching it. I thank the students of the course for their patience in testing an unfinished textbook. I would especially like to thank my incredible team of teaching assistants who read and commented on many parts of the book from the perspective of the students.

I want to thank the Carl Gans Foundation for a generous grant that funded the preparation of the illustrations for the book. I want to thank the Society for Integrative and Comparative Biology (SICB) and the Society for Invertebrate Morphology for being my intellectual homes-away-from-home for many years. Most of the people who read and commented on chapters are people I met and interacted with at the conferences of these two societies.

Ian Sherman from Oxford University Press provided an incredible amount of help and support. It was only because of his encouragement and enthusiasm that I embarked on this project. Ian is set to retire shortly before the publication of the book. I wish him the best of luck in the new phase of his life. Charlie Bath and Katie Lakina served as project editors and provided important technical support and advice.

Finally, I want to thank the people I spend most of my time with: My lab members and my family. They were the ones who had to deal with my long absences and lack of availability over the past few years. The period of writing the book happened to overlap with a term in a senior academic administrative position (as well as much of the Covid-19 crisis). The juxtaposition of these two meant that I could not devote as much time as I should have both at home and in the lab. I apologize to my family and to my lab members and thank them for their patience.

Ariel D. Chipman
March 2023
Jerusalem

Glossary

abdomen In insects, the posterior **tagma**, usually bearing no walking appendages, and containing most of the reproductive and digestive systems. The term is used inconsistently for the posterior tagma in crustaceans as well. More generally, the region containing the digestive system.

abdominal cavity In vertebrates, the part of the coelom containing the digestive system. Also called the peritoneal cavity.

aboral In radially symmetrical organisms, the pole opposite to the **oral** (mouth) pole.

acoelomate Lacking a **coelom** or secondary body cavity.

actin A structural protein common in all eukaryotes. Has roles in the supportive structures of the cell and in muscle activity.

action potential An electrical signal generated within a **neuron** and transferred along its **axon** to a neighboring cell.

afferent Leading toward. Used in **circulatory systems** for vessels leading toward the heart or gas exchange organ, and in nervous systems for nerve fibers leading toward the **central nervous system**.

aggregative multicellularity A form of multicellularity in which cells that are not genetically identical come together to form a multicellular organism.

allantois In vertebrates, a cavity or sac in the egg of amniotes, which serves as a receptacle for waste.

allometric Changing at different rates with a change in general size. Displaying **allometry**.

allometry The tendency of different organs to change at different rates as the organism changes size, due to physical and geometric constraints.

alpha taxonomy The branch of **taxonomy** dedicated to identifying, defining, and describing species.

alternation of generations A life history mode in which generations with differences in form, mode of reproduction, or lifestyle alternate. This may be an alternation of sexual and asexual stages, or of haploid and diploid individuals.

amnion In vertebrates, a fluid-filled cavity that surrounds the embryo and allows it to develop with no connection to environmental water. The defining character of Amniota. In insects, one of the extra-embryonic membranes.

amoebocyte In sponges, an **amoeboid** cell type, which is found within the **mesohyl**. Amoebocytes are motile cells whose main role is digestion of food particles. Can differentiate to form **ova**. Also known as **archeocyte**.

amoeboid Having an amoeba shape. Used to describe a type of motility in single-celled eukaryotes, where cells move through the extension of **pseudopodia**.

ampullae of Lorenzini Organs in the head of some vertebrates that are used to detect electrical fields.

anagenesis An evolutionary process in which a species changes gradually to form a new species, without splitting into two species.

anal pore In ctenophores, transient **aboral** openings that function as the terminal opening of the digestive system.

anamorphic In the context of segmentation, a mode of development in which the organism hatches without the full complement of segments, and additional segments are formed post-embryonically.

antenna (pl. antennae) A long and thin sensory structure at the anterior of an animal. The term is usually used for the sensory structures of mandibulate arthropods, but is used inconsistently in other organisms as well.

antennule The first pair of **antennae** in crustaceans.

anterior–posterior axis The main axis of bilaterian animals, running from the front (anterior) to the back (posterior) end. The longitudinal axis.

anus The posterior opening of the digestive system. The opening through which excess undigested material (**feces**) is removed.

aortic arch In vertebrates, one of the arteries leading from the heart to the gills and from the gills to the **dorsal aorta**.

apical Relating to the tip or apex. In epithelial cells, this is the pole that is distal to the **basement membrane**.

apical tuft A group of hair-like structures at the tip of some types of larvae. Forms the basis of the nervous system.

appendage An extension of the body that has a specific function. In arthropods, most segments bear a pair of appendages that may be used for locomotion, feeding, sensation, or many other functions.

archenteron In embryonic development, the body cavity that forms after **gastrulation** and forms the basis for the digestive system.

archeocyte see amoebocyte.

Aristotle's lantern In echinoid echinoderms, a structure found in the inside of the mouth. Composed of five calcitic teeth and used for processing food.

artery A circulatory vessel that carries blood from the heart toward the body.

article A single unit of a segmented structure. Used to describe segments of arthropod sensory organs or **antennae**.

asconoid A type of body organization in sponges. The simplest type of organization in which pores lead directly into the sponge's central cavity. Also known as ascon.

atrium (pl. atria) A large cavity or space. In tunicates, the space between the body and the tunic. In vertebrates, the region of the heart that blood first enters from veins.

autotrophic Generating energy independently, without relying on energy fixed by other organisms.

axial mesoderm In vertebrate development, a region of mesoderm that lies along the main embryonic anterior-posterior axis. Gives rise to the notochord.

axon A projection from a **neuron** that transfers electrical signals to other cells.

basal Close to the base. In epithelial cells, this is the side close to the basement membrane. In systematics, more basal means closer to the root of the tree.

basement membrane An acellular matrix that is the foundation upon which epithelial tissues are built.

benthic Relating to or living upon or within the sea floor or other surface at the bottom of the water column.

binomial name The official scientific name of an organism. Composed of the genus name (**generic name**) followed by the species name (**specific name**).

biological species concept A concept that sees species as being defined by the ability of members of the species to reproduce with other members of the same species.

biosphere The sum of all environments on Earth that support life, with all the organisms included within them.

biphasic life cycle A life history mode in which adults and juveniles are distinguished both morphologically and ecologically, with a dramatic transition or **metamorphosis** between the juvenile (often known as a **larva**) and the adult.

biramous Having two branches. In arthropod limbs, the two branches are the **exopod** and the **endopod**.

bladder In the excretory system, a sac that stores urine before it is removed from the body.

blastocoel The first embryonic cavity. A liquid filled space within a ball of cells.

blastopore The opening through which cells invaginate throughout the gastrulation process. Forms the beginning of the **archenteron**.

blastula An early stage in embryogenesis, in which the embryo is a hollow ball of cells.

blood The fluid of the **circulatory system**. The term is usually reserved for the fluid within **closed circulatory systems** but is sometimes used more generally.

bone The internal skeleton of vertebrates. The term is used both for individual elements of the skeleton and for the material forming the skeleton:fibrous connective tissue mineralized with calcium phosphate (apatite).

book gills The gas exchange system of marine chelicerates. Made up of a water-filled cavity including individual leaf-like structures where gasses are exchanged.

book lungs The gas exchange system of some arachnids. Derived from the **book gills** of marine chelicerates. An air-filled cavity including individual leaf like structures where gasses are exchanged.

branchial heart A muscular organ that pumps **blood** or **hemolymph** toward the **gills**.

breathing A form of gas exchange where oxygen-rich air is inhaled and carbon dioxide-rich air is exhaled.

breeding colony A group of animals of the same species that come together in a defined space for the purpose of reproducing.

byssus Fibers secreted by some species of bivalve mollusks for the purpose of anchoring them to a hard surface.

caecum (pl. caeca) A blind extension of the gut, usually filled with **endosymbiotic** bacteria.

camera eye A type of eye in which light passes through a small hole or aperture into a spherical space lined internally with photoreceptor cells.

capillary a thin tube in which liquid moves by surface tension. In **closed circulatory systems**, capillaries are the blood vessels that reach all the tissues of the body.

captaculum (pl. captacula) In scaphopod mollusks, thin tentacles used for gathering food and as sensory organs.

carnivory A mode of feeding, where the main source of food is other animals. Animals that feed by carnivory are known as carnivores or **predators**.

cartilage A type of flexible connective tissue made of a gelatinous matrix built upon a scaffold of extracellular fibers. Cartilage is common in vertebrates but is found in other phyla as well.

cell type a group of cells that share a similar structure, function, and gene expression profile.

cell The basic unit of biology. A cell is a membrane-bound entity filled with cytoplasm and **organelles**. All living organisms are composed of cells.

central nervous system A structured collection of neurons and related cells, in which the main information processing of the organism takes place.

cephalization The presence of a head. The process of the evolution of a head.

cephalon The head.

cephalothorax The anterior **tagma** of chelicerates (the **prosoma**) and in some cases, of crustaceans. Contains both sensory and feeding structures, and locomotory structures.

cercaria (pl. cercariae) The early larval stage of parasitic flatworms.

cercus (pl. cerci) Posterior facing sensory appendages in mandibulate arthropods.

cerebellum A part of the vertebrate hindbrain, devoted mostly to motor coordination.

cerebrum A part of the vertebrate forebrain. In mammals, it is the largest and most highly developed region. Responsible for high-level processing.

chaeta (pl. chaetae) Thin bristles found on the **parapodia** or on the body wall of many annelids. Also called **seta**.

chelicera (pl. chelicerae) The anterior appendage pair of chelicerates. Usually function as piercing or cutting mouth parts.

chelifore The anterior appendage of pycnogonids (sea spiders). Homologous to the **chelicerae** of other chelicerates.

chemical building blocks The basic chemical units that make up all organic matter. Include molecules such as sugars, amino acids, fatty acids, nucleotides, sterols, and others.

chemoautotrophy The ability to generate energy through the breakdown of simple inorganic molecules such as methane and sulfur.

chemoreception A **sensory modality** that allows organisms to detect specific chemicals in their environment.

chemosensory An organ or structure devoted to **chemoreception**.

chemotaxis Movement in response to the gradient of a specific chemical. Can be positive (movement toward increasing concentration of the chemical) or negative (moving away from increasing concentration of the chemical).

chitin A complex hydrocarbon, best known as the main component of the cuticle of arthropods, but also found in other animals, as well as in fungi.

chloroplast An organelle in photosynthetic eukaryotes that is responsible for **photosynthesis**. Acquired through the **endosymbiosis** of cyanobacteria.

choana In vertebrates, the connection between the nasal cavity and the pharynx.

choanocyte In sponges, a cell type that is responsible for generating the currents that flow through the sponge body and for capturing food particles that are brought into the sponge's body by these currents.

choanoderm In sponges, the inner cell layer that includes **choanocytes**.

cilium (pl. cilia) Hair-like extensions found in many cells, both in single-celled eukaryotes and in many epithelial cells in metazoans. Used both for locomotion and for generating flow or transferring small particles.

ciliary band A ring of cilia found in the mid-region of several types of larvae.

ciliary motion Locomotion generated by the beating of numerous cilia on the outer surface of the body. Mostly found in small organisms.

circular muscles Muscles that surround the body wall and serve to contract the entire body at a specific point. Circular muscles are also found in the digestive system.

circulatory system The system responsible for delivering dissolved materials to and from tissues of the body. In circulatory systems, a dedicated fluid carrying these dissolved materials is pumped actively through vessels for at least some of its circuit around the body.

circumesophageal ganglion A cluster of neuronal cell bodies surrounding the anterior digestive system. Functions as the brain in many phyla. In some phyla, it is referred to as the circumesophageal commissure or circumesophageal connective.

clade A phylogenetic branch. The basic unit in **phylogenetic systematics**. Equivalent to a **monophyletic group**.

cladistics A branch of systematics that is based on ordering taxa into nested **monophyletic groups**. Also known as **phylogenetic systematics**.

cladogenesis The process of an evolutionary lineages splitting into two lineages.

class In **Linnaean taxonomy**, the level above an order and below a phylum. Mammalia and Insecta are examples of classes.

cleavage The first stage of embryogenesis, in which the zygote goes through a series of synchronized cell divisions to go from one cell to a multicellular embryo.

clitellum In annelids, an enlarged region of the external epithelium, which has a central role in mating and reproduction and ultimately forms a reproductive **cocoon**. The clitellum is the defining character of Clitellata.

clonal colony A group of organisms that are all derived from the asexual reproduction of a single founder individual. In a clonal colony, all individuals are genetically identical.

clonal multicellularity A form of multicellularity in which all cells are derived from the division of a single founder cell. In clonal multicellularity, all cells are genetically identical.

closed circulatory system A type of **transport system** in which **blood** circulates through distinct vessels throughout all of its circuit around the body.

cnidocyte A **cell type** found only in Cnidaria, specialized for injecting venom into other animals. Also known as a **nematocyte**.

cochlea In vertebrates, a **mechanosensory** structure specialized for picking up vibrations at specific frequencies and transferring information about them to the **central nervous system**.

cocoon In clitellates, a structure in which the eggs are laid and where they are protected until they hatch. In some holometabolous insects, a structure in which the animal is protected throughout the process of **metamorphosis**.

coelenteron The main body cavity of **diploblastic** organisms that lack a through-gut.

coelom The secondary body cavity in many bilaterians. Formed within the mesoderm.

coenosarc A tissue connecting different **polyps** in colonial cnidarians.

colloblast A **cell type** found only in ctenophores. Specialized for sticking to and trapping prey items.

colonial Living in a colony, or a group of individual organisms of the same species. Relating to life in a colony.

columnar In epithelial tissue, composed of cells that are taller than they are wide; column shaped.

commissure A cord connecting concentrations of neurons.

community In ecology, a group of organisms of different species that interact within a defined geographical region.

conjugation A form of sexual reproduction involving exchange of genetic material among individuals, found in single-celled eukaryotes.

connective tissue Three-dimensionally arranged tissue that serves to connect other tissue types, fill gaps between tissues, or provide support. In the context of this book the term is usually used as a contrast to epithelial tissues. In the context of histology, the term refers to a more specific type of tissue.

convergence The evolution of similar structures, functions, or characters in unrelated lineages as a result of similar selective pressure.

coprolite Fossilized **feces**.

cornea A transparent layer of tissue covering the eye.

counter-current exchange A mechanism for transferring dissolved molecules or energy between two vessels carrying fluids in opposite directions. See Figure 24.1.

crawling A mode of locomotion in which the ventral surface of an animal is close to the substrate. Used inconsistently in different contexts.

crop An anterior region of the digestive system, where food is stored temporarily. Some preliminary mechanical breakdown of the food may take place in the crop.

cryptobiosis A form of dormancy in which the organism's functions are almost completely suspended. Used in some organisms as a way to survive periods of harsh environmental conditions.

ctene The comb rows of ctenophores. Composed of clusters of cilia that beat in unison.

ctenidium (pl. ctenidia) The comb-like gills of some groups of mollusks.

cuboidal Having a cube shape. In epithelial tissue, composed of cells that are the same height and width.

cup-eye A type of eye in which a pigmented layer of photoreceptive cells is invaginated into the body surface, allowing light to be collected from different directions.

cutaneous respiration Gas exchange through the outer body surface, without the involvement of distinct gills.

dauer larva A life history stage found in some nematodes, in which the organism enters a state of dormancy in response to harsh environmental conditions. Not actually a larval stage according to the common definition.

definitive host The host in which a parasite reproduces sexually. Also known as **primary host**.

dendrite A projection of a **neuron** that serves to transfer electrical signals from other neurons to the cell body.

dermal bone A type of bone that is formed by ossification of dermal tissue.

dermatome The component of the **somite** that gives rise to the **dermis**.

dermis In vertebrates, the connective tissue of the integument that lies under the **epidermis**.

detritivory A mode of feeding where the main source of food is decayed organic matter or detritus. Also spelled detritovory. Animals that feed by detritivory are known as detritivores.

deuterostome An organism in which following **gastrulation**, the **blastopore** becomes the anus.

deutocerebrum The middle region of the tripartite arthropod brain. Also known as the deuterocerebrum.

diaphragm A thick membrane separating two cavities or spaces. In terrestrial vertebrates, the diaphragm separates the abdominal cavity from the pericardial cavity and often becomes muscular to aid in lung ventilation.

dichotomy In phylogenetics, a split of a single branch into two branches.

diencephalon The posterior part of the vertebrate forebrain.

digestive system The system responsible for bringing food into the organism, breaking the food down mechanically and chemically and making its breakdown products available to all the organism's tissues.

diploblast An animal having two **germ layers**. Also known as diploblastic.

diplosegment In diplopods (millipedes), a structure composed of two fused adjacent segments. Each diplosegment normally carries two pairs of appendages.

direct development A life history mode in which the juvenile is morphologically and ecologically similar to the adult, and the juvenile matures gradually to adult form without dramatic metamorphosis.

domain In **Linnaean taxonomy**, the highest level, above a kingdom. There are three domains of life.

dorsal The side of the organism facing upward or away from the substrate. In many bilaterian organisms (with the exception of vertebrates) the dorsal side includes the **circulatory system**.

dorsal aorta In vertebrates, the main circulatory vessel leading **blood** from the gills or from the heart posteriorly toward the body.

dorsal root ganglion One of a series of segmentally arranged clusters of **neurons** dorsal to the vertebrate nerve cord. Derived from the **neural crest**.

ecdysis The process of emerging from the old exoskeleton during growth. The defining character of Ecdysozoa.

ecosystem In ecology, all of the organisms, the interactions between them, and the abiotic factors influencing them, within a defined geographical region.

ectoderm The outermost **germ layer**. Forms very early in development and gives rise to most of the integument and to the nervous system.

ectoparasite An organism that lives on another organism and derives resources from the organism on which it lives (the **host**) while causing it damage.

ectosome In sponges, an outer epithelium shared by a number of canal systems.

efferent Leading away from. Used in **circulatory systems** for vessels leading away the heart or gas exchange organ, and in **nervous systems** for nerve fibers leading away from the **central nervous system.**

electroception A **sensory modality** that allows organisms to detect electrical fields.

electrolocation The use of **electroception** for finding potential prey items.

elytra The hardened forewings of beetles that form a protective covering.

embryogenesis The process of development from a fertilized **zygote** up to hatching or birth.

endobenthic Relating to or living within the sea floor or other substrate at the bottom of the water column.

endochondral bone A type of bone that is formed by ossification of cartilage.

endocrine system The system responsible for chemical communication between different organs of the body. Includes the glands or tissues that synthesize the chemical signals (**hormones**), the tissues that receive the signals and the chemical themselves.

endoderm The innermost **germ layer**. Forms early in development and gives rise to the epithelial tissues of the digestive system. In older literature, sometimes spelled entoderm.

endolecithal cleavage A type of **cleavage** in which cells divide deep within the yolk without the egg itself splitting into separate cells.

endoparasite An organism that lives inside another organism and derives resources from the organism in which it lives (the **host**) while causing it damage.

endopod In arthropods, the main axis of a **biramous** limb. Usually, the portion that is used for locomotion.

endostyle An organ in chordates that produces mucus that leads food into the digestive system.

endosymbiosis A type of interaction between organisms, in which one organism lives inside another organism, for the mutual benefit of both. Endosymbiosis is the accepted explanation for the evolution of mitochondria and chloroplasts.

energy source The origin of the energy used to power metabolic processes within an organism.

ephyra (pl. ephyrae) In cnidaria, a juvenile medusa as it detaches from its mother polyp.

Epibenthic Relating to or living upon the sea floor or other surface at the bottom of the water column.

epidermis The outermost epithelial layer of the integument.

epimorphic In the context of segmentation, a mode of development in which the organism hatches with the full complement of segments.

epithelial cell A very common family of **cell types** characterized by a polar structure, with an **apical** pole and **basal** pole. Epithelial cells connect to other cells laterally to form **epithelia**.

epithelium A common tissue type that forms two-dimensional structures lining cavities or creating borders. Also known as epithelial tissue.

epitoke In annelids, a life history stage in which the organism fills with gametes and changes its behavior in order to reproduce.

erythrocyte A red blood cell. A vertebrate-specific **cell type** that is filled with the oxygen transporter hemoglobin, giving it a red color.

esophagus The section of the digestive system that connects the **pharynx** to the **stomach**.

estivation A type of dormancy that takes place during the summer to protect against desiccation.

eusociality The most complex form of colonial lifestyle. In eusociality, only a small subset of the individuals in the colony reproduce, and all other members contribute to the survival of the colony.

eversible pharynx A structure found in several phyla, in which a portion of the pharynx can be everted to collect food and then internalized to bring the food into the digestive system.

evolutionary arms race The constant improvement and co-evolution over time of characters in different organisms. As organisms evolve mechanisms for improved hunting or competing with organisms of other species, the other species evolve better protective or competitive mechanisms, leading to the evolution of additional improvements in the first species.

evolutionary novelty A character or structure that appears in a species without it being present in its ancestors. An evolutionary novelty is a character that does not have homologs in other species.

evolutionary species concept A concept that sees species as being defined by their shared evolutionary history.

excretory system The system that is responsible for removing waste products from the body and for maintaining a correct balance of water and dissolved ions.

exopod In arthropods, the secondary axis of a **biramous** limb. In many cases it functions as a gas exchange structure or gill.

exoskeleton A support structure that covers the organism from the outside. It is usually part of the integument, or lies above the integument. Also called **external skeleton**.

external gill A gas exchange organ that is formed by an increase in the surface area of the integument in a specific region of the body and is in direct contact with the external environment.

external skeleton See **exoskeleton**.

family In **Linnaean taxonomy**, the level above a genus and below an order. Felidae (cats, tigers, and relatives) and Carabidae (ground beetles) are examples of families.

feather An integument structure found only in birds. A feather is a branched extension of the keratinous layer of the epidermis that has a role in thermoregulation and in flight.

feces The waste products of the digestive system. Undigested material that is ejected through the anus.

filter feeding See **suspension feeding**.

fitness In evolutionary biology, a measure of evolutionary success. Fitness of a character is expressed in terms of the average number of offspring that organisms with this character produce.

flagellum (pl. flagella) A whip-like structure found in many single-celled organisms and used for locomotion.

flame cell A type of excretory cell found in the **protonephridia** of some taxa. Named for its shape that resembles the flame of a match.

foot In Mollusca, the muscular ventral portion of the body, used for locomotion in most molluscan taxa.

forcipule The venom claw of centipedes.

fulcrum In mechanics, the point on which a lever rests and around which it rotates.

furcula The posterior ventral appendage of collembolans (springtails). Functions as a spring.

gamete A reproductive cell. Gametes in most multicellular eukaryotes have half of the genetic complement of other cells. In most organisms, there are male gametes or **sperm** and female gametes or **ova.**

ganglion (pl. ganglia) a cluster of neurons that functions as a processing unit.

gas exchange In animals, the absorption of oxygen from the surroundings and the release of carbon dioxide. Gas exchange is carried out both at the organismic level and at the level of individual tissues.

gas exchange system The system responsible for absorbing oxygen from the surroundings and supplying it to the tissues, and for removing accumulated carbon dioxide from the organism.

gastrodermis In cnidarians, the tissue layer that lines the **gastrovascular cavity** or **coelenteron.**

gastrovascular cavity See **coelenteron.**

gastrulation The developmental process in which external cells invaginate to the interior of the embryo, thus forming the **mesoderm** and defining the embryonic axes.

generic name The first part of the **binomial name**, giving the name of the **genus** to which the organism belongs. The generic name is italicized and capitalized.

genus In **Linnaean taxonomy**, the level above a species and below a family. A genus includes several closely related species.

germ layer One of the three populations of cells that differentiate very early in embryonic development and give rise to a broad range of tissues:**ectoderm, endoderm,** and **mesoderm. Diploblasts** have only two germ layers: **ectoderm** and **endoderm.**

germ layer theory The theory that suggests that all embryonic tissues can be traced to three early **germ layers**.

gill arch One of a series of cartilaginous support structures in vertebrates, originally functioning as gill supports, but modified over evolution to form a range of skeletal elements. Also called **pharyngeal arch.**

gill slit In chordates, one of a series of openings in the pharynx, originally functioning as a mechanism for filtering food particles out of the water, then, in vertebrates, as an opening through which water moved as part of a gas exchange system, and in terrestrial vertebrates becoming reduced or lost. Also called **pharyngeal slits.**

gill A gas exchange organ that is formed by an increase in the surface area of the integument in a specific region of the body. Can be internal or external.

gland A cell or group of cells that produce a specific compound and secrete it.

gnathal segments In mandibulate arthropods, the three segments that carry appendages that function as mouthparts.

gonad The organ that generates and stores **gametes**.

gonopore An opening through which gametes are secreted or absorbed. The reproductive opening.

gonozooid A specialized **zooid** in some groups of cnidarians that carries a reproductive role.

hemocoel A body cavity, derived from the **coelom**, that functions as part of the circulatory system and is filled with hemolymph.

halteres Modified wings in dipteran insects that function as balance organs.

head capsule In arthropods, a rigid structure that encompasses anterior segments and is morphologically distinct from posterior segments.

heart A muscular organ that pumps **blood** or **hemolymph.**

hematopoiesis The developmental process in which blood cells and other cells of the circulatory system differentiate.

hemimetabolous In insects, having an incomplete life cycle that does not include metamorphosis.

hemolymph The fluid of the circulatory system. The term is usually used for the fluid in

open circulatory systems, to distinguish it from **blood**, but the usage is not consistent.

herbivory A mode of feeding in which the main source of food is multicellular plants. Animals that feed through herbivory are known as herbivores.

hermaphrodite An organism that can produce both male and female gametes.

heteronomous segmentation A form of segmentation in which there are regions with different types of segments along the body.

heterotrophic Eating other organisms as a main energy source, thus making use of the energy they fixed.

histolysis Breaking down the organism's own tissues as a source for building new tissues.

holoblastic cleavage A type of **cleavage** in which cell divisions cut through the entire egg.

holometabolous In insects, having a complete life cycle that includes metamorphosis.

holotype A specimen kept in a scientific collection, which is the official representation of the species, and upon which the description of the species is based.

homeothermic Maintaining a constant or almost constant body temperature.

homology Evolutionary sameness. Similarity that is derived from a common ancestor.

homonomous segmentation A form of segmentation in which all segments are similar.

hormone A chemical that is secreted in one part of the body to transfer information to another part of the body.

host The organism that a parasite parasitizes.

hydromedusa A type of **medusa** found in hydrozoan cnidarians.

hydrostatic skeleton A support structure that is composed of an internal cavity filled by an incompressible fluid.

imaginal disc In holometabolous insects, a group of cells that are found in the larva and differentiate during metamorphosis to give rise to adult structures.

imago The adult stage of insects.

immune system The system that is responsible for defending the organism against invasion by pathogens and parasites. Unlike most other systems, it is not usually composed of dedicated organs, but sometimes includes specific cells throughout the body or within the circulatory system.

instar a specific life history stage of an organism with discontinuous growth. Specifically, one of the individual stages between molts in ecdysozoans.

integument The body covering. A complex system that includes several tissue and cell types.

intercalary In insects and myriapods, the reduced segment that lies behind the antennal segment. Homologous to the second antennal segment in crustaceans.

intermediate host A host in which a parasite does not reproduce or reproduces only asexually.

intermediate mesoderm In vertebrates, a region of mesodermal tissue lying between the **paraxial mesoderm** and the **lateral mesoderm**. Gives rise mostly to the **excretory system**.

internal gill A gas exchange organ that is formed by an increase in the surface area of the integument in a specific region of the body and is housed in a compartment that is not in direct contact with the surroundings.

internal skeleton see **endoskeleton**.

interstitial Living in or relating to the space between the particles of the substrate.

intestine The section of the digestive system that is mostly dedicated to absorption of digested food. Often has a twisted or looped shape.

introvert An eversible feeding structure in the head of scalidophorans.

isometric Changing at similar rates with the change in body size. Displaying **isometry**.

isometry The phenomenon of different structures maintaining similar proportion regardless of size.

isotonic A liquid having the same concentration and composition of dissolved ions as the surroundings.

jaw A structural component of the **mouth** of many organisms that is mineralized and helps in preliminary breakdown of food before it is ingested.

joint A meeting point of two skeletal elements that are able to move relative to each other.

keratin A protein commonly found in the outermost layer of the epidermis of many vertebrates.

kingdom In **Linnaean taxonomy**, the level above a phylum and below a domain. Metazoa is a kingdom.

kleptocnidism Use of cnidarian cnidocytes by an organism that feeds on the cnidarian and sequesters its stinging cells for its own use.

labium a feeding appendage formed by the fusion of paired appendages in the third gnathal segment of insects and some myriapods.

labrum An anterior unpaired appendage-like structure in arthropods. Possibly homologous to the antennae of onychophorans and fossil arthropod relatives.

larva (pl. larvae) The juvenile stage of many marine organisms. Usually very different from the adult, and often planktonic. The term is sometimes used for immature stages of terrestrial arthropods.

lateral line A **mechanosensory** structure found in primitively aquatic vertebrates. Composed of an elongated depression that is in contact with the surrounding water and contains **neuromasts**.

lateral plate In vertebrates, a paired embryonic mesodermal structure, that gives rise to many adult structures, including the walls of the **coelom**.

lecithotrophic A larva that feeds on yolk supplied by the mother during the formation of the egg.

leuconoid A type of body organization in sponges. The most complex type of organization in which pores lead into canals that lie under a thickened body wall and connect to all regions of the body. Also known as leucon.

lever In mechanics, a simple machine composed of a rigid structure that rests on a point, known as a **fulcrum**.

Linnaean taxonomy A system of organizing organisms into a series of hierarchical levels.

lobopodium (pl. lobopodia) Unsegmented appendages found in extant and fossil relatives of arthropods.

locomotion Movement that acts to transfer an organism from one place to another.

locomotory system The system responsible for locomotion. Includes muscles, skeletal elements, appendages, and elements of the integument.

longitudinal muscles Muscles that connect along the anterior–posterior axis of an organism or of a structure, such as the digestive system. Contraction of these muscles shortens the main axis or bends the axis, if they contract only on one side.

lungs A gas exchange organ composed of an air filled cavity connected to the surrounding air by a narrow opening.

macromere One of a group of larger cells that are the outcome of unequal **cleavage**.

macromolecule Any one of a number of large molecules that have a role in biological functions, including proteins, nucleic acids, complex carbohydrates, and others.

macrovory A mode of feeding in which the main source of food is large organisms, either plants or animals. Usually used in contrast with **suspension feeding** or **detritivory**. An animal that feeds by macrovory is known as a macrovore.

madreporite In echinoderms, a specialized perforated skeletal plate that allows movement of water from the surroundings into the **water vascular system**.

Malpighian tubule The excretory organ of insects and myriapods. Probably evolved convergently in the two groups.

mammary gland An organ that secretes a fatty nutritious liquid (milk) that female mammals use to feed their offspring. One of the defining characters of mammals.

mandible In mandibulate arthropods, the main feeding appendage, located on the first gnathal segment. In vertebrates, one of the bones of the lower jaw.

mantle In mollusks, an epithelial organ that secretes the shell.

mantle cavity In mollusks, a space underneath the mantle, usually in the posterior (except for in gastropods). The digestive system and the reproductive system empty into the mantle cavity. In many cases, gas exchange occurs in the mantle cavity.

marsupium A pocket in which a female broods the young. Found (convergently) in isopod crustaceans, in some fish, in marsupial mammals, and others.

mass extinction An event in which extinction levels in the entire **biosphere** are significantly higher than background extinction levels.

maxilla (pl. maxillae) In mandibulate arthropods, the feeding appendage located on the second gnathal segment. In vertebrates, one of the bones of the upper jaw.

maxilliped In crustaceans, appendages posterior to the gnathal appendages that aid in collecting food and bringing it to the mouth. In centipedes, an alternative name for the **forcipule** or venom claw. Also spelled maxillipede.

mechanoreception A **sensory modality** that allows organisms to detect mechanical and physical forces acting upon them. Includes touch detectors, motion and gravity detectors, air and water pressure detectors, and others.

mechanosensory An organ or structure devoted to **mechanoreception**.

medulla oblongata The posteriormost region of the vertebrate brain. Connects without a clear border to the dorsal nerve cord. Responsible for most autonomous body functions.

medusa A life history stage in cnidarians that is mobile and responsible for sexual reproduction.

meroblastic cleavage A type of **cleavage** in which cell divisions do not cut through the entire egg and an undivided yolk region remains below the developing embryo.

mesencephalon The embryonic midbrain of vertebrates.

mesenchymal cell A cell that forms a loose and unorganized tissue type, mesenchyme.

mesenterion In vertebrates, a membranous structure, derived from a fusion of the two sections of the **lateral plate**, that connects organs suspended in the coelom to the rest of the body.

mesoderm The intermediate **germ layer**. Forms during **gastrulation** and gives rise to most connective and support tissue.

mesoglea The acellular intermediate tissue of **diploblastic** animals.

mesohyl The intermediate collagenous tissue of sponges.

mesonephros In vertebrates, the adult excretory organ of non-amniotes. Appears as a transient embryonic structure in amniotes.

metacercaria (pl. metacercariae) One of the late larval stages of parasitic flatworms.

metamerism An organization of organs or structures in serially repeated units along the anterior-posterior axis. Also known as metameric organization.

metamorphosis A dramatic change in body shape and function that often forms the transition between the larval stage and the adult.

metanephridium (pl. metanephridia) The multicellular excretory organ of several phyla.

metanephros The adult excretory organ of amniote vertebrates.

micromere One of a group of smaller cells that are the outcome of unequal **cleavage**.

microvillus (pl. microvilli) Small hair-like structures on the apical surface of epithelial cells.

miracidium (pl. miracidia) One of the early larval stages of parasitic flatworms.

mitochondrion (pl. mitochondria) The energy producing organelle of eukaryote cells.

mobile A lifestyle in which an organism is able to actively move from place. Also known as motile.

modular Composed of distinct units that can be arranged or modified individually.

molting The process of emerging from the exoskeleton during growth and replacing it with a new larger exoskeleton.

monophyletic group A group of organisms containing a common ancestor and all its descendants.

morbidity Damage caused to an organism as a result of an attack by a parasite or pathogen.

morphological species concept A concept that sees species as being defined by their shared morphological characters.

morula An early stage of embryonic development, where following **cleavage** the embryo is composed of a cluster of cells resembling a berry.

motile see **mobile**.

motility The degree to which an organism is capable of independent locomotion.

mouth The anterior end of the digestive system. The opening through which food is ingested.

movement Any change of position of one part of an organism relative to other parts.

muscle A type of contractile tissue that drives movement of organs or structures.

mutable collagenous tissue In echinoderms, a unique type of connective tissue that can change its physical properties in response to neural stimuli.

myelin sheath A layer of fatty material that covers the **axon**, contributing to its insulation and improving the transfer of electrical signals along it.

myocytes The contractile cells that make up **muscle** tissue.

myosin A protein that is the main component of **muscles**.

myotome The component of the **somite** that gives rise to segmental muscles.

natural selection The mechanism driving evolution. The tendency of traits that increase an organism's chance of surviving and reproducing to become more common in a population.

nauplius (pl. nauplii) The larval stage of many crustaceans.

negative chemotaxis A tendency to move away from increasing concentration of a certain chemical.

nektonic Actively swimming in the water column.

nematocyst The organelle within a **cnidocyte** that houses the barb and venom that are injected into another animal.

nematocyte see **cnidocyte**.

neodermis The integument of parasitic flatworms. The defining character of Neodermata.

nephron In vertebrates, the individual excretory units that make up the kidney.

nervous system The system responsible for integrating information gathered from the different sensory structures and translating it into responses. Composed mostly of **neurons** and related cells.

neural crest A vertebrate-specific embryonic tissue. Originates in the margins of the **neural plate** and migrates throughout the body to give rise to a range of tissues.

neural plate In vertebrates, a dorsal thickening of the ectoderm that will sink to give rise to the **neural tube**.

neural tube In vertebrates, the embryonic structure that will give rise to the hollow nerve cord that is at the basis of the **central nervous system**.

neuromast A **mechanosensory** organ found in vertebrates that detects gravity and acceleration.

neuron The main **cell type** of the **nervous system**. The simplest data integrating unit.

neurula A stage of embryonic development in vertebrates, in which the neural plate forms and starts to sink in and the embryo starts to elongate.

nociception A **sensory modality** that detects damage to the organism. The feeling of pain.

nomenclature A branch of **systematics** that deals with giving names to organism and taxa, and with the rules for giving names.

notochord In chordates, a dorsal mesoderm-derived rod that provides structural support. In vertebrates, the notochord is a transient embryonic structure.

nucleus The organelle in eukaryotic cells that contains the DNA and is responsible for its regulation.

nymph A larval stage of hemimetabolous insects. Used inconsistently in other arthropod taxa.

ocellum (pl. ocelli) A simple eye with a lens that is found in several phyla.

ocular Relating to the eyes. In arthropods, the anteriormost segment, bearing the eyes.

olfactory cell A **cell type** that is the main component of the **chemosensory** system in many animals. Functions in detecting volatile chemicals.

ommatidium (pl. ommatidia) The basic unit of a compound eye in arthropods.

ontogeny The development of an organism from fertilization to death. Includes both **embryogenesis** and post-embryonic development.

open circulatory system A type of **transport system** in which **hemolymph** circulates through distinct vessels throughout part of the circuit and through the **hemocoel** for part of the circuit.

opisthaptor A specialized attachment organ found in monogenean parasitic flatworms.

opisthosoma The posterior body **tagma** of chelicerates.

oral Relating to the mouth. In radially symmetrical animals, the side in which the mouth is located.

oral–aboral axis The main body axis of radially symmetrical animals.

order In **Linnaean taxonomy**, the level above a family and below a class. Carnivora (carnivorous mammals) and Coleoptera (beetles) are examples of orders.

organelle One of the subcellular structures that together allow a **cell** to function.

organism The basic unit of biology. An organism is a biological entity that uses an energy source and simple chemical compounds from the environment in order to grow and to create copies of itself, based on chemically encoded information.

organizer In developmental biology, a group of cells that produce a chemical signal that provides information for patterning of other cells.

organogenesis The process late in **embryogenesis** in which organs start to differentiate and attain their final structure.

organ a biological structure composed of different tissues and carrying out a specific function.

osculum (pl. oscula) In sponges, the main opening through which water exits the sponge body.

osmoconformer An organism that matches the concentration of ions in its body to that of the surroundings.

osmoregulation The process of regulating the concentration of ions in an organism's internal fluids. Also, being an **osmoregulator**.

osmoregulator An organism that maintains a constant concentration of ions in its internal fluids.

osphradium (pl. osphradia) An ectodermal sensory organ in molluscs.

ostium (pl. ostia) In sponges, one of the pores through which water enters the sponge body.

oviger An appendage found in sea spider males, used to carry eggs.

ovum (pl. ova) The female gamete.

ozopore The openings of cuticle glands in millipedes that secrete unpleasant chemicals to deter potential predators.

palp An elongated sensory structure, usually arising from the head, that has a mostly **mechanosensory** function, but may also include a **chemosensory** function.

parallelism Independent evolution of similar structures in related organisms, using similar cells or developmental mechanisms.

paraphyletic grouping A group of organisms that includes a common ancestor, but not all of its descendants.

parapodium (pl. parapodia) In annelids, lateral extensions of the body wall that may function as paddles for swimming, appendages for crawling, or gills for gas exchange.

parasite An organism that lives on or in another organism in a relationship that benefits the parasite but harms the other organism:the **host**.

parasitoid An organism that lives on or in another organism in a relationship that benefits the parasitoid and leads to the death of the other organism:the **host**. Parasitoids often infect the eggs or offspring of the host.

parietal mesoderm In vertebrates, One of the units of the **lateral plate** that gives rise to the body wall musculature, to ventral mesoderm, and to the **circulatory system**.

parsimony The principle that the simplest explanation is most likely to be the correct one. In systematics, a principle for constructing phylogenetic trees so that they include the smallest number of character changes.

pedicellaria (pl. pedicellariae) In echinoderms, pincer-like structures derived from the mesodermal skeleton.

pedipalp The second appendage pair of chelicerates. Have a wide variety of roles in different chelicerate taxa.

pelagic Living within the water column.

pereon The middle tagma of several crustacean taxa. Roughly equivalent to the insect **thorax**.

pericardial cavity In vertebrates, the region of the coelom that surrounds the heart and lungs or gills.

pericardial coelom A cavity derived from the coelom that surrounds the heart in some phyla.

peripheral nervous system The components of the nervous system that are outside the **central nervous system**. Mostly involved in sensory reception.

peristaltic locomotion A type of **locomotion** found in some groups of worms, which involves coordinated and synchronized contraction of **circular muscles** and **longitudinal muscles**.

peristomium In annelids, the region that includes the mouth and accessory feeding organs.

phagocytosis The process in which a cell engulfs a food particle.

pharyngeal arch see gill arch.

pharyngeal slit see gill slit.

pharynx The first region of the digestive system, immediately following the mouth. Often has a role in preliminary processing of the food.

pheromone A chemical secreted by an organism in order to transfer information to other organisms, usually of the same species.

photoreceptor cell A **cell type** that is capable of detecting light. The basic unit of all photoreceptor organs.

photosynthetic An organism that uses light to synthesize energy-rich organic compounds.

phototrophic generating energy independently by using light to synthesize organic compounds.

phylogenetic systematics A branch of **systematics** that aims to arrange organisms based on their evolutionary history and evolutionary relatedness.

phylogenetic tree A graphical representation of the evolutionary relatedness among organisms.

phylogeny The evolutionary history and relatedness of a group of organisms.

phylotypic stage A stage in **embryogenesis** that is similar among all members of a phylum.

phylum (pl. phyla) In **Linnaean taxonomy**, the level above a class and below a kingdom. Often characterized by a conserved body plan. Chordata and Arthropoda are phyla.

pigmentation The coloration of the integument.

pinacocyte In sponges, a cell type that forms the external epithelial covering of the body.

pinacoderm In sponges, the cell layer that includes the **pinacocytes**.

placenta In mammals, a structure composed of both maternal and embryonic tissue, which serves to transfer nutrients from the mother to the embryo. Placenta-like structures are found convergently in several taxa outside of mammals.

planktonic Living in the water column and being mostly moved by water currents rather than by independent locomotion.

planktotrophic Plankton eating. Often used to describe marine larvae that feed on plankton before settling.

planula The flat early larva of cnidarians.

pleon The posterior tagma of some crustaceans. Roughly equivalent to the insect abdomen.

pleurite The paired lateral **exoskeleton** plate of arthropods.

pluteus The larva of sea urchins.

polar capsule A structure found in myxozoan cnidarians that is probably homologous to the **cnidocyte**.

polyp A life history stage in cnidarians that is usually **sessile**. In Anthozoa it is the only stage and can reproduce both sexually and asexually. In Medusozoa, it usually reproduces asexually and generates **medusae**.

polytomy In phylogenetics, the split of a branch into more than two daughter branches. Usually indicates insufficient data.

population In ecology, a group of organisms from the same species that share the same geographical region and reproduce freely.

positive chemotaxis A tendency to move toward increasing concentration of a certain chemical.

pre-gnathal segments The three anterior segments of arthropods. Probably represent the ancestral arthropod head.

predator An animal that feeds mostly on other animals, usually hunting them while they are still alive.

primary host see **definitive host**.

primary urine The liquid that is initially filtered by the excretory system, before it is concentrated.

prokaryote An organism whose cells do not contain a nucleus or membrane-bound organelles.

pronephros In vertebrates, the first embryonic excretory organs. Functions in larval amphibians, but is transient in all other vertebrates.

proprioception A **sensory modality** that allows organisms to sense the location and position of their own organs.

prosencephalon The embryonic forebrain of vertebrates.

prosoma The anterior **tagma** of chelicerates.

prostomium In annelids, the anteriormost non-segmental region of the body.

protist A single celled **eukaryote**. Considered an obsolete term, but still found in some older literature.

protocerebrum The anteriormost region of the tripartite brain of arthropods.

protonephridium (pl. protonephridia) The single-celled excretory organ of several phyla.

protonymphon The larva of sea spiders.

protostome An organism in which following **gastrulation**, the **blastopore** becomes the mouth.

proximal causality The immediate and direct reason for an event or an adaptation.

pseudocoelom A secondary body cavity that is lined with mesoderm on the outside and is in contact with the endoderm of the digestive system on the inside.

pseudopodium (pl. pseudopodia) An extension of an **amoeboid** cell which is used either for prey capture by **phagocytosis** or for movement.

pupa In **holometabolous** insects, the life history stage in which the organism becomes dormant and undergoes metamorphosis.

pygidium In annelids, the posteriormost region of the body. Also used inconsistently to describe the posterior or tail region in other organisms.

radial cleavage A type of **cleavage** in which at the eight-cell stage all of the cells are arranged in a cube, with fully overlapping cell layers.

radial symmetry A type of symmetry in which body structures repeat following rotation of the organism around a rotational axis.

radula A mineralized feeding organ found in mollusks.

rectum A posterior region of the **digestive system**, usually just anterior to the anus. Functions mostly to reabsorb excess fluids from the **feces**.

reproductive isolation The final stage of the speciation process. A stage at which members of the two daughter species can no longer reproduce with each other.

reproductive system The system responsible for creating and transferring gametes for fertilization with gametes of another individual of the species. Contains the **gonads** and the reproductive organs.

respiration The process of absorbing oxygen and releasing carbon dioxide.

respiratory system See **gas exchange system**.

rhombencephalon The embryonic hindbrain of vertebrates.

rhombomeres In vertebrates, repeated structures within the hindbrain.

rhopalium (pl. rhopalia) In cnidarians, structures that contain a concentration of sensory organs.

scala naturae The idea that all of nature can be arranged along a ladder of increasing complexity. This idea is largely discredited today.

scalid Small spines or scales that are typical for Scalidophora.

sclerotome The portion of the **somite** that gives rise to segmental bone tissue, including most of the vertebrae and some of the ribs.

scolex An anterior structure in tapeworms, used to connect the parasite to the intestinal epithelium of the host.

secondary simplification The process of a structure or character becoming simplified relative to an ancestral more complex condition.

secondary urine The concentrated product of the excretory system that is ultimately removed from the body.

sedentary A lifestyle mode in which an organism is capable of locomotion, but moves very little or very slowly.

segmentation A type of body organization in which an organism is composed of repeating complex units along the anterior–posterior axis of the body.

selective advantage A characteristic that improves an organism's chance of surviving and reproducing.

selective pressure An external influence that increases the chance of certain adaptations becoming fixed in a population or species.

sensory modality A source of information about the world that can be detected by dedicated sense organs.

sensory placode In vertebrates, an embryonic thickening of the ectoderm that interacts with mesoderm to give rise to sense organs.

sensory system The system responsible for detecting information about the world via different **sensory modalities**.

septum transversum In vertebrates, the membrane that separates the **abdominal cavity** from the **pericardial cavity**.

serpentine locomotion A type of locomotion in which longitudinal muscles contract in alternation on both sides of the body, leading to wave like arrangement of the body. Also called **sinusoidal locomotion**.

sessile A lifestyle mode in which an organism is fixed to the substrate and does not move at all.

seta (pl. setae) see **chaeta**.

sexual selection Selection that favors structures or behaviors that increase an organism's chance of finding a mate.

shell A mineralized protective structure that is external to the organism.

simple life cycle A life history mode in which juveniles develop gradually into adults without a dramatic transition.

single-celled algae One of several types of photosynthetic, single-celled eukaryotes. Not a real taxonomic group.

sinusoidal locomotion See **serpentine locomotion**.

siphon A tube that allows water in or out of an organism. Found in several **suspension feeding** taxa.

skull fenestration An evolutionary process in amniote vertebrates, in which the weight of the skull was reduced through the appearance of windows between the skull bones. The position of these windows (fenestrae) is specific to different amniote lineages.

smooth muscle A type of muscle found mostly in the wall of digestive systems. Smooth muscle fibers tend to be shorter than those in **striated muscle**, and they are arranged in sheets.

soaring A type of aerial locomotion which involves relying on air currents rather than on active flapping of the wings.

solenocyte An excretory **cell type** found in **protonephridia** of some animals.

solitary Living as single individuals rather than in a **colony**.

soma In **neurons**, the cell body. The main site of data integration. In reproductive biology, the tissues that are not involved in sexual reproduction. The **somatic cells**.

somatic cell A cell that includes two copies of the genetic material and is not involved in sexual reproduction.

somite In vertebrates, an embryonic cluster of cells that gives rise to most mesodermal segmental structures.

somitogenesis The developmental process of **somite** formation.

speciation The evolutionary process of a single species splitting into two species.

species The fundamental unit of evolutionary biology. A population of organisms that are similar and share genetic material. The lowest rank in **Linnaean taxonomy**.

species concept A way of formalizing **species** and defining their borders.

specific name The second part of the **binomial name**, giving the name of the **species** to which the organism belongs. The specific name is italicized but not capitalized.

sperm The male gamete. Normally a small flagellated cell.

spermatheca An organ in the female **reproductive system** of many animal taxa, used to store **sperm** before it fertilizes the **ova**.

spermatophore A package in which **sperm** is delivered to the female. Found in many animal taxa.

sphaeridium (pl. sphaeridia) A type of balance organ found in sea urchins. Probably derived from spines.

spinneret An organ found in spiders that is used to spin silk. Derived from appendages of the **opisthosoma**.

spiracle An opening in the body wall. Usually, part of the **gas exchange system**, and connected to interior respiratory organs.

spiral cleavage A type of **cleavage** in which at the eight-cell stage there are two layers of cells offset from each other by 45°.

spirocyte A type of **cnidocyte** found in some cnidarians, specialized for clinging to prey.

spongin A structural protein found in the connective tissue of sponges.

spongocoel The central cavity of some sponges.

squamosal Scale like. In **epithelial tissues**, having cells that are wider than they are tall.

statocyst A type of **mechanosensory** organ, found in many animal taxa, that detects gravity and acceleration.

statolith A small crystal, embedded in a liquid or gelatinous environment, that provides the inertial mass for a few types of **mechanosensory** organs.

stereom A type of mineral organization of calcium carbonate found in the skeleton of echinoderms.

sternite The ventral skeletal plate of the arthropod **exoskeleton**.

stomach The section of the digestive system that is mostly dedicated to chemical breakdown of food. Often has a sac-like shape.

stone canal In echinoderms, a mineral-lined canal connecting the **water-vascular system** to the surrounding environment.

striated muscle A type of muscle, mostly dedicated to body **movement** and **locomotion**. Arranged in long **syncytial** bundles.

strobilation In cnidarians, the process of generating juvenile **medusae** by transverse splitting of a **polyp**.

subdermal cavity In sponges, the space just below the outer covering of the body.

suspension feeding A type of feeding in which the main source of food is small particles of organic matter or small organisms, suspended in the water column and collected via a filtering or sieving organ.

syconoid A type of body organization in sponges. An intermediate level of organization in which pores lead into finger-like chambers around a central chamber. Also known as sycon.

syncytial A type of tissue organization, in which individual cells are not separated by membranes, and the tissue functions as a large cell with multiple nuclei.

syndetome The portion of the **somite** that gives rise to tendons.

system A collection of organs and cells that work together to provide a central organismic function.

systematics The branch of biology devoted to questions of organismic diversity and relatedness.

tagma (pl. tagmata) One of the functional body units in arthropods.

tagmatization The organization of the body into functional units or **tagmata**.

taxon (pl. taxa) A group of organisms with shared characters and a common ancestor.

taxonomy The branch of **systematics** devoted to defining characters of different taxa, naming them, and arranging them hierarchically.

teat In mammals, an external structure through which milk from the **mammary gland** is fed to the offspring.

telencephalon The anteriormost region of the vertebrate embryonic forebrain. Gives rise to the **cerebrum**.

teleological An argument that assumes the reason for a phenomenon is its ultimate outcome. Teleological thinking is contrary to the way evolution works.

telson A posterior non-segmented structure in segmented animals.

tentacle a flexible, elongated appendage, usually used for finding and catching food, and bringing it to the mouth.

tergite The dorsal skeletal plate of the arthropod **exoskeleton**.

terrestrialization The evolutionary process of transition from the aquatic realm to the terrestrial or aerial realm.

test An external support structure made by the fusion of skeletal elements.

thermoregulation The ability to control or regulate the internal temperature of the organism.

thorax In insects, the middle **tagma**, carrying the locomotory appendages.

through gut A digestive system with anterior and posterior openings.

tissue An arrangement of cells of one or a few **cell types** that together perform a specific biological function.

torsion In gastropod mollusks, the rotation of the **visceral mass** by 180° relative to the **foot**.

trachea (pl. tracheae) A gas exchange organ, found in several terrestrial animals, mostly arthropods. Composed of branching tubes that penetrate the body and connect to the surrounding air.

transport system The system responsible for transporting dissolved gasses, nutrients, and waste to and from the tissues of the body.

tri-radiate pharynx A triangular organization of the **pharynx** found in some members of Ecdysozoa.

triploblast Composed of three **germ layers**.

tritocerebrum The posteriormost region of the arthropod brain.

trochophore A type of larva found in several phyla within Spiralia. Characterized by a median ciliary band and an apical tuft of cilia.

tube foot A echinoderm-specific structure that is part of the **water vascular system**. Tube feet are membrane-bound extensions that pass through the body wall into the surrounding water.

type specimen A specimen kept in a scientific collection, which contributed to the description of the species.

typhlosole A dorsal expansion of the intestinal epithelium, found in earthworms.

ultimate causality The deep underlying reason for an event or an adaptation.

uniramous Having a single branch. In arthropod limbs, this is the type of limb found in terrestrial taxa.

Urbilateria The hypothetical common ancestor of all bilaterian animals.

urine The product of the excretory system that is removed from the body.

valvular swimming A type of **locomotion** found in a small number of molluskan bivalves, in which opening and closing of the valves generates a water jet that moves the animal.

vein A blood vessel leading to the **heart**.

veliger A type of **larva** found in marine snails.

ventral The side of the organism facing downward toward the substrate. In many bilaterian organisms (with the exception of vertebrates) the ventral side includes the **central nervous system**.

ventral aorta In vertebrates, the blood vessel leading from the heart to the gills.

ventricle A fluid filled cavity. In the vertebrate brain, one of four cavities that form during early development of the brain. In the vertebrate heart, the muscular portion that pushes the blood out of the heart.

vertebra (pl. vertebrae) One of the bones composing the vertebral column, which is the main axial skeleton of vertebrates and gives them their name.

vestibular system The system responsible for balance and orientation. Includes one or more **mechanosensory** organs that detect gravity and acceleration.

villus (pl. villi) Thin protrusion of the intestinal epithelium, composed of one or more cells.

visceral mass In mollusks, the dorsal portion of the body that contains the digestive system, reproductive system and most of the hemocoel.

visceral mesoderm In vertebrates, one of the units of the **lateral plate** that gives rise to the mesodermal components of the digestive system.

water vascular system A system found only in echinoderms that carries out diverse functions including excretion, osmoregulation, gas-exchange, and locomotion. Composed of a series of internal water-filled tubes and external **tube feet**.

worm A general term for long and thin bilaterally symmetrical animals with no appendages. Not a valid taxonomic group.

zooid In colonial animals, a single animal within the colony.

CHAPTER 1

The hierarchical nature of biology

Hierarchical complexity

The biological world is complex, and organisms are complex entities. This is a basic fact we must accept as a starting point for any discussion of organismic biology. We try to simplify and organize biology in order to make it easier to understand (and easier to teach). One of the factors that make it possible to simplify biology is that biological complexity is not chaotic. It is hierarchical and nested. "Hierarchical" means that there are different levels or units of complexity, each of which includes within it the complexity of the lower levels. "Nested" means that any biological unit can be a member of only one unit at the next level of complexity. To understand this idea, let's start by exploring the levels of complexity from the lowest biological level to the level of the organism—the focus of this book (Table 1.1).

The smallest unit in biology is the biological **macromolecule**. Smaller units are in the realm of organic chemistry. This is of course an arbitrary starting point, which not all biologists will necessarily accept. For the purpose of this exploration, it will suffice. Macromolecules include proteins, nucleic acids, polysaccharides, and other complex molecules that can only be formed by biological processes and are only found as components of biological entities. They are usually dealt with in the context of biochemistry or molecular biology.

Macromolecules are often assembled in complex ways to form subcellular **organelles**. There are several types of organelles. Some are bounded by membranes, such as the nucleus or the Golgi apparatus, while others are unbounded complexes of macromolecules like the ribosomes. These are the basic components of **cells** and their arrangement and structure distinguish different types of cells.

There are hundreds or even thousands of cell types, including muscle cells, neurons, bone cells, epithelial cells, and numerous others. Organelles and cells are studied within the realm of cell biology.

Many cells of the same type or of a few types are organized together in a specific spatial organization to give rise to **tissues**. Tissues will often fulfill a certain biological function, such as secretion from glandular tissues, mechanical protection in epithelial tissues, or contraction in muscle tissues. However, tissues usually do not function alone, and several tissue types work together to give functional **organs**. For example, the skin is an organ that includes epithelial tissues (the epidermis), glandular tissues (sweat and fat glands), and connective tissues (the dermis). Tissues and their arrangement within organs are studied by the discipline known as histology.

Different organs function together as a **system**. The organs of a system can be connected, as in the

Table 1.1 Examples of hierarchical levels of biological complexity

Organizational level	Example	Function
System	Digestive system	Providing nourishment to the organism
Organ	Stomach	Breaking down food items into usable components
Tissue	Epithelium of the stomach	Lining the lumen of the stomach and secreting digestive enzymes
Cell	Epithelial secretory cell	Synthesizing digestive enzymes
Organelle	Golgi apparatus	Shuttling enzymes out of the cell
Macromolecule	Peptidase	Cutting peptide bonds in proteins found in food items

Organismic Animal Biology. Ariel D. Chipman, Oxford University Press. © Ariel D. Chipman (2024). DOI: 10.1093/oso/9780192893581.003.0001

different organs that make up the digestive system, or they can be distributed throughout the body as in the organs of the sensory system. There is a limited number of organ systems, most of which will be covered in this book. While there are differences in their structure and the organs of which they are made, the general function of the different organ systems is conserved across many types of organisms. The function of individual systems and the organs they are made up of is the focus of physiology. How the different organs systems work together in a functional organism is the subject of organismic biology, and exploring the similarities and differences among the different organ systems is one of the main themes of this book.

The hierarchical organization described above is an example of the way we can make biological complexity easier to understand. Scientists might disagree about where to draw the boundary between levels of complexity and how to define each level, but all in all, the hierarchical view is a realistic representation of how complexity is structured in biology. Nonetheless, it is also inherently imprecise. We have taken a continuum of increasing complexity and broken it down into somewhat arbitrary levels, and have even gone so far as to attribute studies of different levels to different disciplines. This is common practice and is generally useful for organizing biology into manageable problems that are easy to study. The main objective of this book is to deal with manageable generalities and with overarching principles. However, generalities are limited in scope and there is a limit to how far principles can be true. It is important to remember that the world is usually more complex than the simple rules we try to impose upon it.

Complexity above the level of the organism

We have described a hierarchical nested complexity in which each level of complexity is built up of component units, which are in turn complex and made up of their own component units. This suggests that the whole functioning organism is a biological entity of incredible complexity. However, biological complexity doesn't end there. Organisms interact with their biotic and abiotic environment, eat other organisms and are eaten by them, and compete with other organisms for resources.

Just as we can break down an organism into constituent parts, we can use individual organisms as building blocks for larger and ever more complex ecological structures: **populations**, **communities**, **ecosystems**, and the **biosphere**. The biosphere includes different ecosystems, which in turn are composed of interactions among different communities, made up of populations of organisms from different species. Levels of organization above the organisms are dealt with in the science of ecology.

A question that arises when discussing biological complexity is to what extent we need an understanding of all levels of underlying complexity to deal with a higher level. Do community ecologists need to understand what tissues make up the stomach of a bird they are studying? Do population biologists have to know the structure of the cells making up the lizard's muscles? The obvious answer is of course that no one scientist can have detailed knowledge of all levels of complexity, and this caveat is becoming more and more extreme the more our understanding of different levels of biology increases, and the more knowledge is gained about more organisms. The less obvious answer is that a good scientist is at least *aware* of the levels of complexity above and below. Cell biology done with an understanding of the organism from which the cells come, will be more cautious and realistic than cell biology that treats the cell as though it exists in a biological vacuum. A scientist studying the physiology of the nervous system of an organism should be aware of the ecological context in which the animal lives and know what other organisms it interacts with.

Taxonomical hierarchies

In addition to the hierarchical complexity within biological entities, we can also look at the hierarchical relationships among these entities, which are a reflection of their relatedness and evolutionary history. Different organisms are related to a greater or lesser degree depending on the time that has passed since they shared an ancestor. The study of the diversity of organisms and how they are related is known as **systematics**. **Taxonomy** is the biological discipline within systematics that defines the characteristics of different organisms, gives the organisms names, and arranges them into higher hierarchical levels.

Taxonomy provides a tool for classifying organisms into hierarchical groups that have a series of characteristics in common. The taxonomy that is most widely used in biology is known as **Linnaean taxonomy** after the eighteenth century Swedish naturalist Carl Linnaeus, who first developed it. The basic unit of Linnaean taxonomy is the **species**, which includes all of the individuals of a distinct type of organism. Note that the word "species" is both singular and plural. Chapter 2 is devoted to the definition and importance of species in biology and, as we shall see there, this is not a trivial definition. Several similar species are grouped together in a **genus** (plural **genera**). Several genera that have unique characters in common are grouped together in a **family**. Similar families are grouped together into an **order**. The taxonomical level above an order is a **class**, and above that is a **phylum** (plural **phyla**). The phylum is the taxonomical level that will be at the focus of this book, with chapters devoted to individual phyla (or sometime a few phyla). A **kingdom** groups together several phyla. After a few introductory chapters, this book will focus on a single kingdom, the animal kingdom or Animalia. The entirety of biological life is divided into three **domains**, each encompassing several kingdoms. We will briefly discuss these domains in Chapter 3. Each one of these levels, from species to domain, forms a **taxon** (plural **taxa**), meaning a grouping of organisms that includes all of the lower-level taxa within it.

The conventions of naming organisms and higher taxonomic groups are formalized under the discipline of biological **nomenclature**. According to these conventions, taxonomic names are always capitalized. Common names in English are often anglicized forms of the taxonomic names, which tend to be based on Greek or Latin . Thus, we can talk about "Arthropoda" or "the arthropods," with the former referring to the taxon and the latter referring to the organisms included within the taxon. Note that a taxonomic name is a proper noun and never has a definite article ("the"), just as we refer to John or Jane and not "the John" or "the Jane." We

Table 1.2 An example of hierarchical Linnaean taxonomy, demonstrated for the house cat

Taxonomic level	Example	Description
Domain	Eukaryota	All organisms made up of cells containing nuclei and complex organelles
Kingdom	Animalia	All animals
Phylum	Chordata	Animals with a dorsal supportive rod
Class	Mammalia	Animals with fur who lactate to feed their young
Order	Carnivora	Carnivorous mammals
Family	Felidae	The cat family
Genus	*Felis*	Small cats
Species	*Felis domesticus*	The house cat

will use taxonomic names (e.g. "Arthropoda") and common names (e.g. "the arthropods") interchangeably in this book, depending on the context.

Linnaean taxonomy is very useful for describing diversity and for discussing the characteristics of individual taxa. However, it does not explicitly reflect evolutionary relationships, as all taxa of the same level within a shared higher-level taxon, are officially considered to be equivalent. This is partially remedied by the introduction of intermediate level taxa. For example, a family can be divided into several subfamilies, and related orders can be grouped together into a superorder, but there is a limit to the resolution this can produce. Many biologists prefer a framework known as **phylogenetic systematics** to describe relationships. Nonetheless, for the purposes of this introductory-level book, we will stay with traditional Linnaean terminology.

Generalizations in biology

Comparative biological research aims to find generalities that hold true in many cases. It looks for common organizational principles and common laws. This is possible, since all biology is based on a shared set of components, and because organisms function within a set of common physical laws. Thus, we can find commonalities (as well as differences) in the way distantly related organisms move, or in the way different organisms deal with similar food sources. Other commonalities stem from shared evolutionary history. Closely related organisms share morphological characters or behaviors because they inherited them from a shared ancestor.

Generalizations are useful for studying and teaching organismic biology, because they save us the trouble of discussing every single organism separately. We can discuss an entire taxon through generalizations that are true for that taxon, and we can discuss a biological process through generalizations that are true for organisms that carry out that process. This book is based on generalizations. We will cover general organizational principles, general functional principles, and general characteristics of major taxa. However, we must keep in mind throughout reading this book, that they are only generalizations, and reality is often much more complex.

While generalizations are useful, the beauty of biology lies in the specific variations that make individual taxa special and different from other taxa. For every rule we describe, there are bound to be exceptions, and often it is the exceptions that are the most interesting. You will notice that the words "usually," "most," and "often" are very frequently used in this book. That is because every generalization has its exceptions, and there is almost no statement we can make in biology that is always true. Comparative biology aims to strike a balance between the generalizations, the variations, and the exceptions, to best understand how organisms function and what factors influence their interactions with the world they live in.

CHAPTER 2

Species concepts and speciation

What is a species?

In our discussion of biological hierarchy, we identified the species as the basic taxonomic unit. We now must tackle the question that arises from this identification: what is a species? A good starting point is to use the following deceptively simple statement: "A species is a group of individual organisms that are all the same." In other words, organisms that share enough characteristics belong to the same species. As an example, all golden eagles or all European hedgehogs look the same and share a series of characteristics and therefore belong to the same species. Clearly, this definition is far from satisfactory and raises more questions than it answers. Indeed, in many cases males and females of the same species, or juveniles and adults, can be very different from each other, despite belonging to the same species. Nonetheless, this statement provides a starting point from which to embark on an exploration of one of the most important, yet most elusive and difficult to define concepts in biology.

Species concepts

According to various counts, there are something between 25 and 30 different **species concepts.** Each of these concepts proposes a way to formalize the "sameness" alluded to above. Some species concepts are abstract and aim to explain what species are and why they exist, whereas others are practical and aim to provide tools for identifying and defining species. Species concepts attempt to contribute to answering two related but fundamentally different questions. The first is "what unites individual organisms belonging to the same species?" and the second is "what separates organisms belonging to different species?"

Let's start with the best-known, most commonly used, and most intuitive species concept, **the biological species concept**. This concept, first formalized by the great evolutionary biologist Ernst Mayr in the 1940s, states that a species is a group of actually or potentially interbreeding natural populations, reproductively isolated from other such groups. Thus, the most important feature defining members of the same species is their ability to interbreed successfully and produce viable offspring. Note that this concept says nothing about the "sameness" of the members of a species. The biological species concept answers both questions presented above: what unites members of a species is a shared gene pool and what separates them from other species is their inability to share genes. We will call the inability to share genes **reproductive isolation,** and as we shall see, this is one of the most important elements of the entire species debate.

A very different species concept is the **evolutionary species concept** formalized by several paleontologists and evolutionary biologists in the second half of the twentieth century, most notably George Gaylord Simpson. This concept sees each species as an independently evolving lineage of ancestor–descendant populations, which maintains its distinction from other such lineages. The main difference between this concept and the biological species concept is that it includes an element of time and continuity, whereas the biological species concept focuses on a fixed point in time. What unites members of a species according to this concept is shared ancestry, and what separates them from other species is independent history.

The final species concept we will discuss (and, as mentioned above, there are many others) is the

Organismic Animal Biology. Ariel D. Chipman, Oxford University Press. © Ariel D. Chipman (2024). DOI: 10.1093/oso/9780192893581.003.0002

morphological species concept. This concept unites members of a species based on shared morphological characters and differentiates them from other species based on differences in these characters. The morphological species concept is the closest to the simplistic idea of "sameness" we started with. It is the easiest for the layperson to understand and the simplest to explain. This concept makes it possible to clearly define an individual species based on an explicit list of characters. Indeed, this is what taxonomists do—they define species based on explicit characters. However, unlike the other two concepts presented above, it is very subjective and leaves open the question of what level of similarity or difference is required to define or differentiate a species. Some species can be very polymorphic (appear in different forms) and deciding what degree of variability is acceptable within a species can be difficult. These problems are best exemplified by a common (though tongue-in-cheek) formulation of the morphological species concept that says that a good species is one that was described by a qualified taxonomist.

Species names

There are currently around 1.5 million named species of animals and about half a million named species of plants, fungi, and unicellular organisms. A named species is one for which there is a scientific name, and that has been formally described in a scientific publication. For every described species there is (or should be) a single individual specimen, known as the **holotype**. The holotype is the individual upon which the description of the species is based. It is explicitly defined in the scientific publication that first described the species and is (or should be) stored in a scientific collection somewhere in the world—usually a national museum or a university collection. Additional specimens can be allied with the holotype and given special status. These are known collectively as **type specimens**. Estimates as to the total number of species on the planet (named and unnamed) vary widely. A reliable average estimate is something around 10 million species.

Every named species has a **binomial name**. This is the genus name followed by the species name. Convention dictates that binomial names are written in *italic script* with the genus (or **generic name**) starting with a capital letter and the species (or **specific name**) with a lowercase letter. Thus, the binomial name of our own species is *Homo sapiens*, while the domesticated dog is *Canis familiaris*. In the full version, the scientific name, that is the binomial name, is followed by the authority—the name of the person who first described the species or who most recently redefined it—and the year of description. When a species is redefined, the authority appears in parentheses. The rules for naming and describing species are complex and very formalized, and we shall not go into them here. The rules regarding naming of animals are overseen by a body known as the International Committee for Zoological Nomenclature (ICZN), with other taxonomic groups having their own equivalent committees.

The reason for all this formalism with holotypes, official descriptions, and naming regulations is to avoid duplicate naming (meaning two species having the same name or the same species having two names), and to circumvent the many problems that arise from the basic difficulties in the idea of species discussed above.

The scientific discipline charged with dealing with species is the science of taxonomy, and the branch devoted to identifying, defining, and describing species is known as **alpha taxonomy**. Taxonomists usually work in the collections or museums already mentioned as the repositories for holotypes. However, these collection house not only type specimens, but a full representation of the biodiversity within their geographical area (some collections aim to represent global and not just regional biodiversity). Thus, every species on earth may be represented by specimens in numerous collections, and each of these collections may have numerous individual specimens of any one species. These collections, some of which include many decades or even centuries of sampling, are an invaluable asset for understanding species distribution, variability within species, or historical changes in biodiversity. They are the resource taxonomists return to periodically to test the validity of known species or to revise their definitions.

How do species change?

Up to now, we've dealt with species as fixed entities. However, species undergo change over time. The idea of species as immutable units was common in pre-modern science and was accepted as a main paradigm until the early nineteenth century. The person credited with challenging this idea is the French naturalist Jean Baptiste de Lamarck (although he was not the first or only one to suggest similar ideas, he is the best-known). Lamarck believed in change through internal driving forces, with a general metaphysical trend for species to improve all the time and climb up an idealized ladder of nature (or *scala naturae*). Half a century after Lamarck, the concept of species changing over time was picked up by Charles Darwin, the father of evolutionary theory, who expanded it and gave it a simple explanation, grounded in basic principles.

Darwin's theory of **natural selection** formed the basis of his ground-breaking book *On the Origin of Species*, published in 1859. The theory is built upon three postulates: 1) Species are made up of individuals who vary in diverse characters or traits; 2) The variable characters are heritable; 3) The differences among individuals affect their chances of surviving and of reproducing, with some characters having a **selective advantage** over others. Taking these postulates together, the theory of natural selection states that individuals with advantageous characters will on average have more offspring, and that these offspring will display the same characters that allowed their parents to survive and reproduce. Over time, these characters will become more common among individuals of the species, until eventually they will be found in the entire species and will be added to its defining characters.

The important point to remember regarding natural selection is that it is individual members of the species that display certain characters, and it is individuals who survive and reproduce (or don't), but the resulting change is in the population and

ultimately in the whole species. Going back to the biological species concept, the fact that members of a species all have the ability to interbreed and have a shared gene pool is what allows characters that confer a selective advantage to spread through the species.

Species change through advantageous characters becoming more and more common and ultimately becoming fixed. The opportunity for change can be the appearance of a novel character through random genetic changes (mutations). This novel character allows the individuals that exhibit it to utilize resources, escape predators and other dangers, or find a mate and reproduce better than their conspecifics who do not exhibit the character. Alternatively, a change in the environment—a new resource or a new danger—causes an existing character to provide an advantage to those who exhibit it, when previously it had no effect on the chance of survival. Either way, a character that was once rare or non-existent becomes common to the entire species.

The mode of change described in the preceding few paragraphs assumes a change sweeping through the entire species, affecting all populations equally. Such a process that causes the entire species to undergo a shift in characters is known as **anagenesis**. However, sometimes changes affect only part of the species, causing a species to split into distinct populations that are different from each other.

The process of speciation

When change affects only part of the species, we start to see a differentiation of distinct populations. Initially, these distinct populations are still members of the same species, since they have the potential to reproduce and are not sufficiently differentiated morphologically to warrant a definition as a new species. However, over time, differences between populations can accumulate to such an extent that they are dissimilar enough to justify being called individual species. Such a splitting is referred to as **cladogenesis** or **speciation**: the process of one species becoming two species.

Speciation can occur when two distinct populations, with limited interaction and interbreeding, are subject to different **selective pressures**. For example, one population may be at a higher altitude, where there is an advantage to having physiological features that deal with lower temperatures. Alternatively, one population may have access to a resource not found in the environment of the other, or be threatened by a predator found only in its environment. In such a case, each population will undergo a gradual shift in relevant characters, possibly in different directions, until the populations are distinct. Speciation can also occur when two populations are isolated enough so that a novel character appearing in one population is not transmitted to the other population. In such a case, one population will remain unchanged, whereas the other will undergo a slow shift. Of course, we can imagine any number of combinations and versions of these scenarios. What all speciation processes have in common is that they end with two (or more) populations that undergo separate evolutionary changes to an extent that they are different enough to no longer belong to the same species (Figure 2.1).

What then are the criteria for deciding whether two populations have differentiated enough to warrant being defined as two species? The answer

to that question is exactly what we discussed at the beginning of this chapter and depends on the choice of species concept. Commonly, the process of speciation is considered to have been completed when the two diverging populations attain reproductive isolation. Under reproductive isolation, the two populations no longer interbreed and no longer share a gene pool.

Populations can be reproductively isolated through a number of different mechanisms. They can undergo changes that prevent males from one population from fertilizing females from the other. They can undergo changes that make females from one population less attractive to males from the other (or vice versa). They can also undergo genetic or molecular changes that make the genetic material from one population incompatible with that of the other (genetic isolation). Reproductive isolation does not have to be absolute. It is enough that members of one population prefer members of the same population for breeding or that offspring of mating between the two populations have some disadvantage for the two populations to be effectively isolated. Indeed, one of the known conundrums of defining species stems from the fact that even reproductively isolated species can sometimes interbreed. In some cases, full isolation

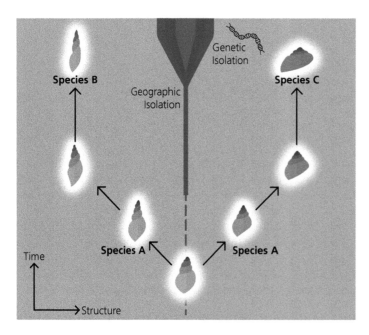

Figure 2.1 The process of speciation: Speciation begins when a population splits in two, and the two subpopulations begin to change independently. Geographic isolation (indicated by a thick blue line—a river) prevents the two subpopulations from mixing, and the differences between them accumulate. Eventually, the two subpopulations are different enough to be genetically isolated, leading to the formation of two distinct species.

takes a very long time to achieve, during which there can be stable hybrid zones between two otherwise distinct species.

Species are dynamic entities

Given the many problems, complexities, and conundrums discussed above, a question that often comes up is whether the idea of a species is useful, and whether in fact species are real biological entities and not artificial constructs set up by taxonomists.

As a general answer, it is probably safe to say that species are real entities, and they are in fact the basic building blocks of biodiversity. However, this answer must always be taken with a pinch of salt, since species don't always behave in the way we expect them to, based on our chosen species concept. We can expect a fairly large proportion of species to be "good species," in that they are morphologically different from members of other species, are reproductively isolated from them, and have their own evolutionary history. We must nonetheless keep in mind that this is not always the case (see Box 2.1). Species are dynamic entities that change over time. We usually see them as a snapshot in time of an ongoing process. At least some of the species we study will be somewhere within the ongoing process of speciation. They may be morphologically distinct from their sister species but not yet reproductively isolated, or they may have achieved reproductive isolation with only a minimal degree of morphological differentiation. Many of the problems, complexities, and conundrums in defining species stem from the dynamic nature of species and from our tendency to want nature to fit into tidy categories.

To summarize; species are real biological entities, but are fluid and complex. How we define and conceptualize species depends on the type of question we are asking, and can vary among disciplines within biology. Regardless of the definition and species concept we choose, any research endeavor in any biological discipline must start by clarifying the species being used for the study, and by making sure that we can identify it and differentiate it from other species.

Box 2.1 Examples that defy classification

Even if it was possible to agree upon a clear and uniform definition of species, there are some cases where it is impossible to determine species boundaries and define species unequivocally. We'll briefly mention two such cases, mostly as an illustration of the limitations of species theory.

Ring species: Sometimes we find a series of populations situated around a barrier so that they form a ring. Each one of these populations is slightly different from its adjacent population, but can freely interbreed with them so they are not considered separate species under any definition or concept. However, there is a point in the ring where two adjacent populations cannot interbreed. These populations are then considered separate species, but they are connected around the ring by a series of population which are conspecific at each point. The two best known examples of ring species are lungless salamanders of the genus *Ensatina* around the Central Valley in California, and gulls of the genus *Larus* around the globe.

Fossil species (or paleospecies): In cases where we have a very good and continuous fossil record, we can follow gradual changes in a species over time. As we move toward later fossils, they will be different enough from earlier fossils to be considered separate species. However, if the change is uniform and gradual (as is sometimes the case), there is no way to draw a line separating the ancestral species from the more recent one. Paradoxically, the poorer the fossil record, the less of a problem we have. If there are no intermediate forms, delimiting the two species is much easier.

CHAPTER 3

What is an organism?

The simplest organisms

What constitutes a biological organism?

The title of this book is *Organismic Animal Biology*. It focuses on the level of the organism, starting with the simplest organisms and adding increasing levels of complexity. Before we can do this, we have to address the question of what an organism is. The definition of an organism is closely tied with the definition of life—a notoriously complex scientific and philosophical debate. The question "What is Life?" is also the title of a 1944 book by the quantum physicist Erwin Schrödinger (also known for his paradoxical thought experiment involving a cat that is both dead and alive). In *What is Life?* Schrödinger attempted to give a physical and chemical basis for the complex phenomenon of biological life, thereby laying the theoretical foundation for the molecular biology revolution that took place over the following decade. Following Schrödinger's influence, we will use a very simple and all-encompassing chemistry-based definition: A biological **organism** is an entity that uses an energy source and simple chemical compounds from the environment in order to grow and to create copies of itself, based on chemically encoded information.

Like all definitions, it is possible to find borderline cases that don't exactly fit—viruses being the most obvious example. Our simple definition also ignores many important aspects of life that are found in more complex and specific definitions. We will come back to these at the end of the chapter. For now, we will use this as a starting point to outline the most general requirements for organisms.

What does an organism need?

Based on our definition, the most fundamental resources an organism needs are an **energy source** and **chemical building blocks**. The latter include the carbon-based constituents of complex macromolecules; sugars, amino acids, carbohydrates, and other molecules, which are described at length in textbooks of organic chemistry and biochemistry. Much of what drives the evolution of different organisms is selection for getting better and more efficient at obtaining these building blocks from available sources by using diverse strategies. The exact identity and relative importance of the required building blocks varies between organisms.

Organisms can be divided into two main types, based on the source of the energy utilized to exploit the building blocks. Organisms that are **heterotrophic** gain energy by breaking down organic molecules and using the energy stored within them. Heterotrophs thus rely on other organisms both for the molecular building blocks and for the energy needed to maintain life. They obtain both of these by feeding on other organisms. All animals are heterotrophs, whether they feed on other animals, on plants, or on the remains of either.

In contrast, **autotrophic** organisms obtain their energy from nonbiological sources. The most common form of autotrophy is **phototrophy**—using sunlight as an energy source for biological functions. Phototrophic organisms are also known as **photosynthetic** organisms. Nearly all plants are photosynthetic autotrophic organisms, obtaining all of their energetic needs from the sun. A rarer form

Organismic Animal Biology. Ariel D. Chipman, Oxford University Press. © Ariel D. Chipman (2024). DOI: 10.1093/oso/9780192893581.003.0003

of autotrophy is **chemoautotrophy**. These organisms use simple inorganic molecules as their energy source. These molecules, most commonly methane, hydrogen sulfide, and ammonia, are in a reduced chemical state, and are high in energy. Oxidizing these compounds provides enough energy to supply the organism's needs. Chemoautotrophy is found almost exclusively in Bacteria and Archaea (see below), often in environments that are too harsh for other organisms.

Autotrophic organisms usually (but not always) are also able to synthesize most of the basic building blocks needed for their growth. Thus, the resources they require are only simple inorganic molecules. Indeed, the original source of nearly all organic building blocks is photosynthetic organisms. After these molecules are first synthesized, they move through the food chain until they eventually break down.

What is the simplest organism?

With these basic definitions in mind, let's add a few more requirements to identify the simplest biological organisms. An organism needs to be separate from its environment in order to maintain its individuality and to function. Therefore, with very few exceptions, all organisms on Earth are cellular. That is to say, they are all made up of cells that are separated from the environment by a cellular membrane. All organisms on Earth have double-stranded DNA as their chemically encoded source of information (viruses are exceptions to this rule, but we have already excluded them from our definition of organisms).

The simplest true organisms are therefore single cells, surrounded by a membrane, and including within them DNA and the molecular machinery that can take materials absorbed from the environment and use them for growth. This is exactly a description of the organisms known as **prokaryotes**. Included within this group are Bacteria and Archaea, two of the three domains of life. Discussion of prokaryotes is beyond the scope of this book, and the interested reader is referred to texts on microbiology.

The third domain of life, Eukaryota, is also largely made up of unicellular organisms. The cells of eukaryotes are significantly more complex that those of prokaryotes, and tend to be much larger. Because multicellular organisms, including animals, emerged from within unicellular eukaryotes, we will survey the diversity of this group briefly, before discussing multicellularity in Chapter 5.

The unicellular eukaryotes

Eukaryotes are characterized by cells that have internal membranes separating the different organelles. Most notably, they have a membrane-bound **nucleus** that contains the cell's genetic material—the DNA. Eukaryotes can be heterotrophic or autotrophic. Autotrophic eukaryotes are exclusively photosynthetic, and the capture of light to generate energy is done within specialized organelles known as **chloroplasts**. Another important type of specialized organelles is the **mitochondria**, which are responsible for aerobic metabolism—the process of using oxygen to break down sugars into carbon dioxide and water, while releasing usable energy.

The reason these two types of organelles are worth mentioning specifically is because they are central to our understanding of the evolution and function of the eukaryotic cell. It is generally accepted that both mitochondria and chloroplasts derive from ancient prokaryotic bacteria, which were taken up by an ancient proto-eukaryotic cells in a process known as **endosymbiosis** (Figure 3.1). According to this evolutionary model, iron-reducing bacteria were taken up once, in an ancestor of all current eukaryotes, and gradually became integrated into the cells as mitochondria, taking on many of the cell's metabolic roles. In a later event, an ancestor of photosynthetic eukaryotes (already with mitochondria) took up photosynthetic bacteria related to today's cyanobacteria, and these allowed the cell to generate energy using sunlight, becoming integrated into the cell as chloroplasts. It is not clear whether this type of event happened only once, or whether it occurred several times in different lineages. What is clearer is that there have been several cases of secondary endosymbiosis, where a cell that already had chloroplasts was engulfed by another cell. The resultant cell, an ancestor of several groups of

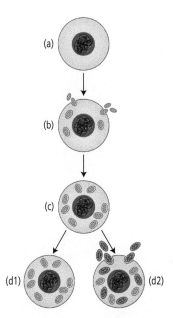

Figure 3.1 The process of endosymbiosis: The proto-eukaryotic cell (a) probably had a differentiated nucleus. During the first endosymbiotic event (b) iron-reducing bacteria were taken up by the cell, leading to the ancestor of all extant eukaryotes (c) with a nucleus and mitochondria. In one lineage (d2), a second endosymbiotic event occurred, with cyanobacteria entering the cell to form chloroplasts. This is the ancestor of all photosynthetic eukaryotes. Other lineages (d1) never incorporated chloroplasts. These include animals, fungi, and many other eukaryotic groups.

Amoeboid   Ciliate   Flagellate

Figure 3.2 Classification of single-celled eukaryotes is based on movement type. Amoeboid eukaryotes move by extending temporary appendages or pseudopodia. Ciliate eukaryotes move by coordinated beating of numerous small cilia. Flagellate eukaryotes move by the waving of one or two long flagella.

It is important to stress that these are mostly functional classifications and are generally not valid taxonomic groups. The following section covers a sample of some of the best-known and most significant groups of unicellular eukaryotes, but it is far from being complete.

The diversity of unicellular eukaryotes

Photosynthetic protists are traditionally referred to by the blanket term of **single-celled algae**. This type of organism is found within many different taxonomic groups. The best known are probably Euglenida, a diverse group of freshwater algae, which are often responsible for the green color of standing water in pools or puddles. Euglenids move using a long, whiplike structure known as a **flagellum** (plural **flagella**) and can thus be classified as flagellated autotrophic protists. They have no mineral skeleton, but their body shape is maintained as a rigid structure through an organic cytoskeleton (see Figure 3.3A).

Several important groups of marine algae are worth mentioning, since these organisms are together responsible for a significant portion of primary production (fixing carbon dioxide into organic molecules) in the global ecosystem, and for the cycling of many minerals through the biological food chain.

Coccolithophora—usually spherical with a calcium carbonate skeleton (or **test**) made of a series of round plates. When they die, coccolithophores settle on the bottom of the ocean and their skeletons form the basis of many chalk formations. The famous White Cliffs of Dover are made almost entirely of coccolithophore skeletons.

Diatoms have a silica-based test, often with an intricate structure, and are responsible for close to

eukaryotes living today, has exceptionally complex genetic material, containing remnants of all its different ancestors.

In terms of numbers of named species, the diversity of unicellular eukaryotes is not nearly as high as that of animals. However, in terms of higher level taxonomy, and differences in fundamental cellular characteristics, unicellular eukaryotes are extremely diverse, and are usually classified into at least six *kingdoms*. Some older classifications refer to all unicellular eukaryotes as a single kingdom, known as Protista, but today it is clear that this is not a valid taxonomic ranking. Despite this, **protist** is still used as a colloquial term to cover the many different groups of unicellular eukaryotes.

Eukaryotes can be classified based on several criteria: Energy source (heterotrophy vs. autotrophy), mode of movement (amoeboid, ciliate, or flagellate, see Figure 3.2), type of skeleton (silica, calcium carbonate, or no skeleton), and mode of cell division.

(a) (b) (c) (d)

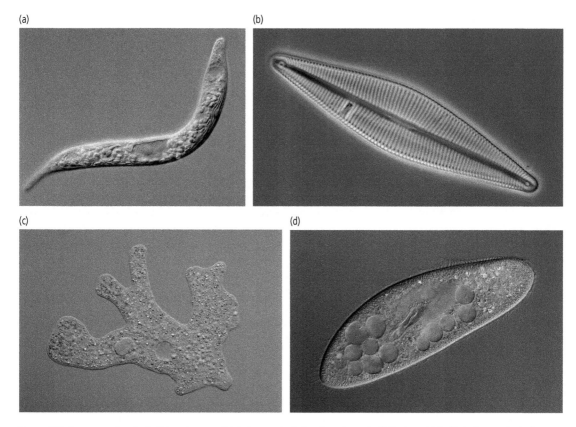

Figure 3.3 Some examples of unicellular eukaryotes: (a) A photosynthetic flagellate, the euglenid *Euglena viridis*. (b) A photosynthetic diatom, unknown species. (c) An amoeba, *Amoeba proteus*. (d) A ciliate, *Paramecium caudatum*

Source: a: Photo supplied from Shutterstock: Maple Ferryman (https://www.shutterstock.com/image-photo/light-micrograph-protist-euglena-mutabilis-showing-1425193058); b: Photo supplied from Shutterstock: Maple Ferryman (https://www.shutterstock.com/image-photo/frustule-pennate-diatom-bacillariphycerae-protista-showing-1962424114); c: Photo supplied from Shutterstock: Lebendkulturen.de (https://www.shutterstock.com/imagephoto/amoeba-proteus-chaos-diffluens-large-broad-261076376); d: Photo supplied from Shutterstock: Lebendkulturen.de (https://www.shutterstock.com/image-photo/paramecium-caudatum-differentialinterference-contrast-dic-100488472).

40% of ocean primary production and almost all of the cycling of silica through the marine food web (see Figure 3.3B).

Dinoflagellata—an exceptionally diverse group of algae. A vast majority are photosynthetic and planktonic in the world's oceans. These usually have an organic test. As their name suggests, locomotion is through a flagellum—often two flagella. Some marine dinoflagellates can go through periodic blooms, when their densities increase to tens of millions of cells per liter. These blooms are known as "red tides" and can be extremely hazardous to other organisms due to toxins produced by the dinoflagellates. In addition to the marine planktonic dinoflagellates, members of this group can also be found as endosymbionts in other marine organisms, such as corals or even other protists (see below). Another group of dinoflagellates lives in the digestive system of terrestrial animals (termites being the best-known example), where they aid in breaking down hard-to-digest compounds.

Amoeboid protists are characterized by an irregular body shape. Their membranes can send out extensions filled with cytoplasm, which are known as **pseudopodia** (singular pseudopod). Locomotion is through extension of pseudopodia, followed by cytoplasm flowing into them, with the entire cell thus "stepping" into the extended pseudopod. Amoeboid protists are all heterotrophic, using their pseudopods to engulf prey and bring it into the cytoplasm where it is digested.

Amoebozoa—the best known example of this type of protists. This is the group that includes the iconic amoebas (or amoebae), which are common in stagnant water (see Figure 3.3C). Some species of amoebozoans cause serious digestive illness.

Foraminifera—heterotrophic amoeboid protists. Many of them have a calcium carbonate **test** and can reach sizes of over a millimeter. Their tests can also be very intricate and complex, and they are sometimes made up of several chambers, looking like minute snail shells. They normally live on the sea floor and do not move. Foraminiferan shells are one of the components of sea sand.

Radiolaria—amoeboid protists with thread-like pseudopodia and an internal silica-based skeleton. They are common planktonic organisms, found mostly in warm and tropical seas. Both foraminiferans and radiolarians occasionally have photosynthetic dinoflagellate endosymbionts, so despite not having chloroplasts of their own, they can get most of their nutritional needs from their endosymbionts, complementing this only occasionally with prey-capture using their pseudopodia.

Many protists are covered in numerous small hair-like locomotory structures known as **cilia** (singular cilium). These protists are collectively referred to as ciliates, and unlike the previous groups we made, they probably are a valid taxonomic group—Ciliata (see Figure 3.3D). The ciliates encompass over 10,000 named species. Although cilia are usually used for locomotion, some ciliates do not move and use their cilia for capturing food. Among the ciliates, the most familiar are species of *Paramecium*, common dwellers of stationary bodies of water. *Paramecium* are large protists, a quarter to a third of a millimeter in length. They are relatively complex for single-celled organisms, with a distinct "mouth," a subdivision into several functional partitions, and two nuclei. While ciliates reproduce by cell division, this is sometimes accompanied by a complex process of exchange of genetic material between individuals, knows as **conjugation**.

Unicellular members of Fungi, the group whose multicellular members include mushrooms, often have significant roles for humans. This group includes the yeast that are used for brewing beer and baking bread, molds that are important for developing the flavor of cheeses, as well as molds that spoil foods. It includes the organisms that cause skin diseases like athlete's foot, ringworm (not a worm at all), and some sexually transmitted diseases. They are also common components of the soil ecosystem, often in association with the roots of plants, and are partially responsible for the unique odors of different soil types.

Several groups of protists include species that are important parasites of humans and are therefore of medical significance. In addition to amoebae mentioned above, there is *Plasmodium*, the agent of malaria, which infects hundreds of millions of people a year, many infections ending in death; *Giardia*, which causes severe intestinal symptoms, also with many millions of cases per year worldwide; Trypanosomes, which cause trypanosomiasis, or sleeping sickness; *Trichomonas*, which causes vaginal infections, as well as others.

More complex needs

Having introduced groups of relatively simple organisms—those that are single-celled—we can go back to our original definition of an organism and its needs and expand it to cover more complex organisms too. All an organism requires are an energy source and basic chemical building blocks, but to obtain these requirements, it needs several functions or systems. An organism needs a system that captures the necessary resources (which for simplicity, we can refer to jointly as "food") and breaks them down into usable components. It needs a system that locates and identifies food, and a parallel system that either brings the food to the organism or brings the organism to the food. In a competitive world, the organism must take care not to become food for another organism, so systems for detecting danger and protecting from danger are also necessary. Finally, no organism can be evolutionarily successful without making additional copies of itself. Thus, some sort of proliferation or reproductive system is crucial for any organism.

All of the functions listed above can be found in unicellular organisms, be they eukaryotes or prokaryotes. As we shall see in subsequent chapters, these functions—and a few others—are found in more highly elaborated form in multicellular organisms. As we go through the different animal taxa and survey the different organ systems found in them, keep in mind that regardless of the com-plexity, all organisms are dealing with the same fundamental requirements and have evolved to ful-fill these requirements in a way that optimizes their chance of reproducing and passing their genes on to further generations.

CHAPTER 4

The concept of evolutionary change

Selection and change

This chapter introduces several key concepts that are used for describing evolutionary changes and their dynamics. As we saw in Chapter 2, species are the fundamental units of evolution. Species are more or less distinct entities that undergo changes over time. The changes are mostly a result of natural selection: the differential survival and reproduction of individuals with different characters. Thus, the key component of evolutionary change is the existence of variability within a population composed of members of the same species.

Organisms live in a complex and challenging world. Natural selection leads to improvement of the organism's ability to deal with different challenges, increasing its chances of obtaining food, avoiding predators, and surviving long enough to reproduce. These different challenges or **selective pressures** constantly pull the organism in different directions. Some evolutionary changes may seem at first glance to be detrimental to the organism. However, the fact that the individuals that harbor these changes survive and manage to reproduce is evidence that under certain circumstances or in conjunction with other characters, they must offer a selective advantage. One well-known example of this is the presence of extreme ornamentation in members of one sex of a species (usually the male). The antlers of deer, the extreme tails of peacocks, or the flashy colors of some fish may make the individuals that present them more susceptible to predation or they may require more energy to maintain. Nevertheless, they also increase the individual's chances of attracting a female and reproducing, and thus, in the balance, confer a selective advantage. This phenomenon is known as **sexual selection**.

Natural selection results in organisms that have optimized their response to different pressures. The outcome of evolution is not the best possible organism in terms of any given criterion or character, but an organism that is a compromise between multiple selective pressures, interactions between characters, the organism's evolutionary history, and the limitations of physics and biology. Evolutionary biologists use the term **fitness** to describe the average chance for an organism with a given suite of characters to reproduce. With a fitness of 1, each individual will on average have one offspring. If the fitness is higher than 1 the number of individuals with those characters will increase and the character will become more common in a population. Characters that confer higher fitness will ultimately become fixed in a population, meaning that all individuals in that population will display the character.

The changes imposed by natural selection are genetic and are therefore inherited and passed down lineages. Once an advantageous character appears in a lineage, it is maintained in all descendants of the species where it was first fixed. The character will remain until such time that it no longer confers an advantage under the balance of all selective pressures. Lineages accumulate more and more changes over time in response to new selective pressures or through the appearance of novel advantageous mutations. As they accumulate changes, they diverge from their ancestors. Different species with a common ancestor will diverge from each other and become less and less similar over evolutionary time (although they may converge in some characters, as discussed later in this chapter). Even as species diverge, they retain some of the characters they inherited from their shared

Organismic Animal Biology. Ariel D. Chipman, Oxford University Press. © Ariel D. Chipman (2024). DOI: 10.1093/oso/9780192893581.003.0004

ancestor, providing evidence for their common ancestry.

Phylogenetic trees

The most common and useful way of representing a **phylogeny** or the relationships between different taxa is through a phylogenetic tree (Figure 4.1). A **phylogenetic tree** is a graphical description of the evolutionary history of a series of taxa and the evolutionary relationships between them. Every phylogenetic tree is a hypothesis about a series of branching events, each representing a case of speciation (see Chapter 2), and about the changes in characters along the tree. Since the tree is a hypothesis (actually, a series of hypotheses), we can test the tree as we would test any scientific hypothesis. This is usually done statistically, giving a measure for how reliable the tree is as a whole, or how much support there is for every individual branching point or character change in the tree (See Box 4.1 for more on how trees are integrated into evolutionary thinking).

A tree is composed of tips, connected via a series of nodes. The tips represent the taxa whose relationships we want to describe. The tips are normally species or higher-level taxa, but in some cases can be populations or even individuals. The nodes represent individual speciation events—the splitting of a single lineage into two distinct lineages that no longer have gene flow between them. Although we mentioned that speciation is a gradual process, for the sake of phylogenetic trees, it is considered to be a single point event. At the base of the tree is the root, which represents the original lineage from which all the tips originated. In an ideal tree, all nodes are binary branching points or **dichotomies**; that is, speciation events that resulted in two distinct lineages. In reality, we don't always have enough information to identify every single binary event, and we must draw a node that splits into more than two branches. Such a node is known as a **polytomy**. A tree that includes a polytomy is incomplete, since the polytomy does not provide an evolutionary hypothesis for that part of the tree.

There are two different ways to look at a tree and interpret what it is shows (Figure 4.1A). It can be read bottom-up or top-down. When starting from the bottom, we are moving forward through time and tracing the evolutionary history of the lineage represented in the tree. We start with the root, which is where the first speciation event in the lineage we are following took place. As we move up the tree, we meet a sequence of speciation events, each represented by a node. Two branches come out of each node (if our tree is fully resolved and has no polytomies), each representing one of the two independent lineages that result from the speciation event. Finally, we reach the terminal taxa at the tips, each representing an undifferentiated entity, which is the final outcome of the evolutionary process we followed.

When starting from the top of the tree, we begin with a series of taxa of interest—the tips. Moving down the tree, we move back in time going through ever more distant ancestors, until we reach a node. Each node represents the most recent common ancestor of two lineages or branches that converge at the node. Continuing down the tree, we meet a sequence of nodes, each representing the most recent common ancestor of all the branches above it. Finally, we come to the root, which represents the most recent common ancestor of all taxa in the tree.

Since every node on the tree represents a common ancestor, every node defines a taxonomic group that includes that ancestor and all of its descendants. Such a group is known as a **monophyletic group** or **clade** (Figure 4.1B). There are cases when we will want to refer to a group that does not include all of the descendants of an ancestor. Such a group is known as a **paraphyletic grouping**. In a strict evolutionary context, paraphyletic groupings are not considered to be real taxa. However, they often appear in common usage and are sometimes useful in ecological or biodiversity contexts. An example of a paraphyletic group is the reptiles. The common ancestor of all reptiles is also the ancestor of birds, but birds are not normally counted within the reptiles. Thus, the Linnaean class Reptilia does not include all the descendants of its common ancestor. The convention is that paraphyletic groupings are distinguished in writing by adding an asterisk after their name, as in Reptilia*.

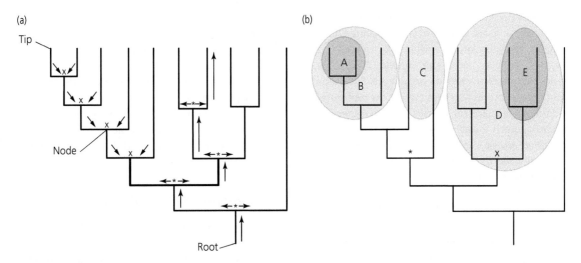

Figure 4.1 Phylogenetic trees: (a) Reading trees: The right side of the tree highlights a bottom-up reading, starting from the root via a series of speciation events (arrows diverging from an asterisk). The left side of the tree highlights a top-down reading, starting from the tips via a series of common ancestors (arrows converging toward an x). (b) Monophyletic and paraphyletic groups: (A) is a monophyletic clade, enclosed within monophyletic clade (B). Similarly, (E) is a monophyletic clade enclosed within (D). The common ancestor of (D) is marked by an x. Conversely, (C) is paraphyletic, since its common ancestor, marked with an asterisk, is also the common ancestor of taxa not included within (C).

Building trees

The process of phylogenetic tree reconstruction is too complex for a detailed discussion here. However it is useful to have a basic understanding of the principles that underlie building trees, to help understand what a tree actually represents. Briefly, the process includes three steps: data collection, tree searching, and estimating tree support.

The first step in tree building is accumulating a dataset of characters for each of the terminal taxa or tips we wish to include in the tree. The dataset can include morphological characters, physiological characters, molecular sequences, or any other type of taxon-specific data. The data are used to build a matrix that includes all of the characters for each one of the taxa, organized in a format that can be read by the tree building tool or software we want to use.

The next step is to look for a tree that best satisfies a series of criteria defined by the tree builder. The simplest criterion is similarity, and a tree built using this criterion would group together taxa that are the most similar or have the most characters in common. This approach is known as clustering. It is now considered to lack statistical power and to neglect the historical aspects of evolution, so is

therefore rarely used. A somewhat more complex criterion looks for the tree that includes the smallest number of character changes throughout the tree (known as the **parsimony** principle). The parsimony principle is at the base of a school of tree building known as **cladistics** that emerged in the 1980s as a revisionary movement in taxonomy in general. Most current studies in phylogenetic reconstruction use much more complex criteria based on statistical models of how morphological characters or molecular sequences change and search for a tree that fits the statistical model best.

Finally, some methods of tree building inherently include an assessment for the statistical support for the entire tree and/or for individual elements of the tree. In other cases, the tree undergoes a support test that involves sampling subsets of the data and testing how well they match the tree generated by the full dataset.

Homology

Homology is one of the most important concepts in evolutionary biology, but it is also one of the most elusive. The roots of the concept of homology are in nineteenth century comparative morphology, when the British anatomist Richard Owen defined homology as "the same organ in different animals under every variety of form and function." To Owen, homologous characters or organs are "the same." Under an evolutionary framework, the definition of homology is historical: characters in two species are homologous if the same character was found in the common ancestor of the two species. The relationship between these two homologous characters is known as homology.

The evolutionary definition of homology is straightforward and fairly simple. If it is so simple, why then is the concept of homology so elusive, and why have numerous heavy tomes being written on the subject? The problem lies in our ability to actually identify homologous characters despite the simple definition, as well as in defining what we mean by the word "same."

Since we rarely know exactly what the common ancestor of two species was like, we are forced to draw conclusions about homology based on comparisons between the characters in two related extant species. There are a number of criteria we can use to do this. Two structures that are in similar location in the body and have similar relationships with other structures are likely to be homologous. If we can break down a character into several constituent characters and find that the constituent characters are similar in the two species, it is likely that the encompassing characters are homologous. Morphological structures that are formed during embryonic development via similar developmental processes and using the same genetic pathways are probably homologous. All the above criteria have to do with similar connections and relationships of putatively homologous structures. We can also use historical or phylogenetic criteria to identify homology, by surveying the presence and absence of a character on a phylogenetic tree. If a character is found in all or most taxa descended from the common ancestor of the species we are looking at, it is likely to be homologous among them all.

Occasionally, we will find a character that is not homologous to any character in any other taxon beyond the one where we first found it. Such characters are said to be examples of **evolutionary novelty**, characters that emerged within a taxon, without being present in its nearest relatives. The question of the origins of novel characters is a subject of intense interest in evolutionary biology. Often, the appearance of a novel character can lead to an evolutionary radiation, resulting in numerous descendant taxa displaying the novelty. How novelties arise is becoming better understood through the science of evolutionary developmental biology, and we will return to this question toward the end of the book (Chapter 32).

Convergence and parallelism

Sometimes, a similar character or structure is found in two species, but we have evidence to suggest that the characters are not homologous, either based on an evolutionary reconstruction, or based on lack on similarity at the level of constituent components. Those characters are said to be the results of convergent or parallel evolution. **Convergence** and **parallelism** occur when two evolutionary lineages are subjected to similar selective pressures. Since all animals are composed of the same biological building blocks and since they all function within the same physical world, examples of convergence and parallelism are fairly common. Organisms in different lineages will follow different evolutionary paths that converge on a similar result. Convergence can be at the level of general body organization, as in the case of the distantly related sharks and dolphins or the "crab" shape in several distinct groups of crustaceans. It can be at the level of a functional structure such as the wings of bats and birds. It can even be at the level of protein structure or molecular interactions. For the purposes of this book, we treat convergence and parallelism together, although there are differences between them, which we will not discuss.

In all these cases, if we try to apply the criteria for the identification of homology, we will find that the criteria are not met. The dorsal fins and tail fins of sharks and dolphins are connected to different structures and have different spatial relationships with other structures in the two lineages. The bones in the wings of bats and birds are different and the muscles attached to them are not homologous. The evolutionary history of the "crab" shape in different crustaceans is different, and it is found scattered across many lineages, with no evidence for a single appearance of the character.

Causality in evolution

A type of question that is often asked (not only in biology) is why things are the way they are. Why do birds have wings? Why do insects have six legs? Questions of this sort can be answered in two ways. We can give an answer that explains the immediate selective advantage of a certain character: Birds have wings to allow them to fly. Insects have six legs because it allows them to walk with an alternating tripod gait. These explanations address **proximal causality**, i.e., the direct and recent cause for the existence of a structure. While these answers are correct, they are partial. A complete answer, especially when dealing with evolutionary questions, has to delve into the underlying deeper causes. This is the **ultimate causality** behind a phenomenon. To give a complete answer that addresses both proximal and ultimate causes, we have to know that birds are descended from bipedal dinosaurs with elongated forelimbs. We have to follow the evolutionary history of arthropods and trace the reduction of limb number down to the six limbs of insects. Perhaps the distinction between proximal and ultimate causality can be better explained using an example from a completely different discipline: history. If we ask what led to the outbreak of World War I, most people will answer that it was the assassination of the Archduke Franz Ferdinand in Sarajevo. This is correct, but it is the proximal cause only. The outbreak of World War I was the ultimate result of a complex series of mutual defense treaties, of imperialist ambitions, and of decades of resentment between the regional powers of Europe, that set the stage for the assassination and the resulting war.

There is an additional problem with the type of answers we gave to the preceding questions: they are **teleological**. That is, they assume that evolution works by finding solutions to problems; by striving to reach a specific goal. In fact, evolution is blind and does not plan ahead. Birds do not have wings because evolution "planned" the best way to allow birds to fly. Wings evolved by a series of miniscule steps, each of which gave the bearers of the

modified wings a better chance of survival and of bearing offspring. At each step along the way, there was variability in forelimb size and shape in the population, with one variant being more successful and ultimately spreading in the entire population, leading to a change in the species.

Finally, a note on "evolutionary shorthand" for the remainder of this book. Repeating this detailed and complex description of selection and variants spreading through the population for every example of evolution is rather tiring. Having explained the principle once, we can be brief in the future. When we say, for example, that birds evolved wings in order to fly, we are abridging the entire evolutionary scenario into one sentence, and we assume the reader understands.

Box 4.1 Tree thinking

A common popular perception of evolution sees it as a linear and progressive process. In fact, evolution includes both linear aspects—the type of changes discussed in this chapter—and branching aspects—the process of speciation and the splitting of evolutionary lineages. Merging these two types of processes in our concept of evolution leads to what has been termed "tree thinking."

Tree thinking is more than just drawing phylogenetic trees. It is an approach to understanding diversity. In tree thinking we are interested in the relationships among organisms, in the evolutionary history that is inherent in these relationships (what was the common ancestor of a group of organisms like), in the way characters are inherited from common ancestors, and in the way these characters change along evolutionary lineages.

This book is explicitly based on a tree-thinking approach to evolutionary biology and to organismic biology. Throughout the book we will come across characters that unite a group of related organisms and characters that are unique to a subset of organisms. Even without drawing a tree, we should be able to conceptualize the notion of "mapping" characters on a tree and thus reconstructing where and when in evolution these characters appeared and where they persist. This type of mapping allows us to differentiate homologous characters from convergent characters and to understand the ultimate causes behind the structure and organization of different animals, and thus to have a complete view of the origins of animal diversity.

CHAPTER 5

Multicellularity

The multiple convergent cases of multicellularity

In Chapter 3 we saw that eukaryotes are very diverse and that their diversity is reflected in multiple aspects of cell size, shape, and function. Our discussion in that chapter was limited to the unicellular members of the domain, which we called protists. However, eukaryote diversity is not limited to protists. In fact, multicellular eukaryotes account for the overwhelming majority of life on Earth, at least in terms of described species number and probably also in terms of biomass (but not in terms of individual organisms, where there is a clear majority for bacteria).

The multicellular animals—Metazoa (or sometimes Animalia)—form the core of this book, and it is therefore most important for us to understand their evolution. However, for this chapter, we will survey different types of multicellularity and different paths to achieving it. Multicellularity has evolved convergently several times within the eukaryotes. Stable macroscopic multicellularity is found within five kingdoms of eukaryotes. The true multicellular eukaryotes include animals, higher plants, red algae, brown algae (kelp), and fungi (mushrooms and their kin, where it has probably appeared several times independently). In addition to these, different degrees of transient or facultative multicellularity have appeared in numerous additional lineages, usually with much lower complexity than that found in the lineages mentioned above. Examples of this simple multicellularity are the cellular slime molds which aggregate to form transient fruiting bodies, the spherical cluster of algal cells known as *Volvox*, chain-forming diatoms, and many others.

What are the advantages of multicellularity?

If multicellularity has evolved so many times, it must have a strong and general selective advantage. One notable advantage is the possibility of increasing in size. As we shall see in Chapter 19, body size is extremely important and has multiple consequences for almost every aspect of an organism's life. Generally speaking, bigger is usually better. Larger size affords better protection against predators, while in turn allowing the organism to eat larger prey (in predatory organisms). It allows more efficient physiological processes and wastes relatively less energy. Since there are physical limitations on the maximum size a single cell can attain, the easiest way to break the size barrier is to evolve into an organism that is composed of multiple cells.

At this point, it is prudent to remind the reader one last time about the "evolutionary shorthand" we adopted in the previous chapter. Of course, evolution didn't plan for the organism to evolve multicellularity in order to enjoy the advantages of size. Individuals that through some molecular change were able to create multicellular clusters had an advantage over those that didn't and, over time, this ability improved through successive small steps, each giving an additional advantage. We simplify this explanation—with the reader's permission—by saying that multicellularity evolved in order to break the size barrier.

Increased size is probably the first and more important advantage to becoming multicellular. However, multicellularity is not only about having multiple cells. The hallmark of complex multicellularity is that the cells are *different* from each other.

Organismic Animal Biology. Ariel D. Chipman, Oxford University Press. © Ariel D. Chipman (2024). DOI: 10.1093/oso/9780192893581.003.0005

A multicellular organism has multiple **cell types**, each with its own shape and function. The diversity of cells allows the distribution of tasks among cells and allows individual cells to specialize in specific tasks. It is this division of labor that confers the most significant long-term advantage to multicellularity, through allowing the fine-tuning of each cell's role, without affecting the roles of other cells. Specialized cells are more efficient at their tasks than generalized cells that must provide all the functions necessary for the organism to survive. The ability for cells to specialize also facilitates the evolution of novel functions that could not exist in a single-celled organism or in an organism with multiple identical (or nearly identical) cells. It is also what allows multicellular organisms to increase in complexity and to diversify, providing the basis for most of the diversity on our planet.

The path to animal multicellularity

There are two main paths through which a single-celled organism can become multicellular. One is through aggregation of multiple cells together to form a colony, known as **aggregative multicellularity**. The other is through cell division without separation of the daughter cells, known as **clonal multicellularity** (Figure 5.1).

Aggregative forms are found in Amoebozoa—the cellular slime molds mentioned above—as well as in some ciliate groups and sporadically in other groups. These aggregations are normally transient and are an adaptive response to stress or reduced resources. In the best-studied cellular slime mold *Dictyostelium*, amoeboid cells come together to form a mobile "slug." This slug is composed of undifferentiated cells that do not feed and do not undergo cell division. The cells move together searching for a suitable environment. When such an environment is found, the slug undergoes a radical transformation, with individual cells differentiating into one of a number of cell types. Some cells form a fruiting body, with the central cells differentiating into spores, while other cells form a supportive structure known as a stalk, and eventually dry up and die. The spores are dispersed by the wind and eventually hatch into amoebae, starting the cycle over again.

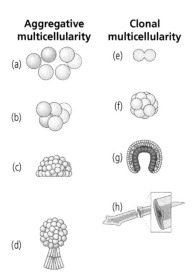

Figure 5.1 Different paths to multicellularity. In aggregative multicellularity, genetically different cells (a) (marked in different colors) come together to form a genetically heterogeneous cluster (b). With time, the cell types differentiate (c) and undergo a morphological rearrangement to give a differentiated structure (d). In clonal multicellularity, a single cell divides (e) to generate a cluster of genetically identical daughter cells (f). These cells then move and undergo differentiation to give a spatially differentiated embryo (g), which ultimately develops to give a mature complex organism (h).

Clonal multicellularity is significantly more common, and all complex multicellular organisms are of this type. Well-studied examples of simple clonal multicellularity are the colonial choanoflagellates (see Box 5.1), which form facultative colonies through the division of a mother cell without the daughter cells separating. The daughter cells maintain cytoplasmic bridges among them, and thus continue to transfer signals and nutrients between the cells. *Volvox* colonies are similar in general structure to some choanoflagellate colonies, despite being composed of very different cells and having a different ecology. Filamentous colonies—long strings of cells connected end to end—are formed by members of several eukaryote taxa through axial cell division without separation.

A question that remains is how complex multicellularity with diverse cell types evolved from simple clonal multicellularity. It has long been thought that differentiation came after the appearance of multicellular forms. According to this scenario, the first step would have been an organism similar to a colonial choanoflagellate or to *Volvox*, with a group of identical cells making up the colony. Over time, the tasks necessary for the colony's survival would be divided among individual cells. The individual cells would specialize to carry out their distinct tasks more efficiently, with additional specialized cell types appearing progressively over time.

An alternative view is based on the fact that many unicellular organisms undergo differentiation over time, with a single individual being able to sequentially transform from one cell type to another. This model for the evolution of complex multicellularity posits that the differentiation modulated over time found in some protists formed the basis for spatial differentiation of complex multicellular organisms. According to this model, the ancestor of complex multicellular organisms was made up of a clonal colony of cells that already had the capacity to differentiate into different cell types. The time-based regulation of differentiation formed the basis for spatially regulated differentiation seen in the more complex descendants. In this model, additional cell types would have evolved over time, gradually becoming optimized toward increasingly specific roles.

Regardless of the path toward multicellularity, there are a number of additional elements needed to complete the transition from single-celled to multicellular organism. There is an entire battery of molecular components involved in holding cells together, in signaling between cells, and in telling cells how to differentiate. The main structural component of multicellular animals is a family of proteins known collectively as collagen. While collagen-like molecules are found in unicellular eukaryotes, only in animals do collagens form complex fibers that provide structural support. Intercellular signaling is based on diffusible proteins from several families, some of which are known to a greater or lesser extent in single-celled eukaryotes. Finally, a battery of small proteins known as **transcription factors** are responsible for activating and repressing different genes in a cell-specific manner to create the differences among different cell types.

Comparative genomics of diverse protists as well as multicellular eukaryotes allows us to trace when and where in evolution these components first appeared and what role they played in the evolution of multicellularity. Not surprisingly, the distribution pattern of these molecular components is complex. Some of them are found in single-celled relatives of multicellular organisms, where they have roles that are not related to multicellularity, but presage them, for example as intraspecific diffusible signals. Other components appear only when multicellularity appears and are thus correlated with its evolution.

Sexual reproduction is a central characteristic of multicellularity

It is fairly clear that sexual reproduction in some form is the primitive state for all complex multicellular organisms. In sexual reproduction, only a subset of cells is involved in transmitting their genes to the next generation (Figure 5.2). The reproductive cells are known as **gametes**, and they are usually found within dedicated structures called **gonads**. Female gametes are known as **ova** (singular **ovum**) and tend to be relatively large. Male gametes are known as **sperm** (singular and plural) and tend to be very small with a flagellum that allows them to swim toward the ovum. In the most common form of sexual reproduction, the gametes contain only one copy of the genetic material of the organism, whereas all other cells, **somatic cells**, have two copies. Male or female gametes are formed in the gonads via meiosis and are then released during reproduction. Male gametes and female gametes of the same species meet and fuse to give rise to a **zygote**, which will go on to develop into a new individual with its genetic material inherited half from the father and half from the mother.

Embryonic development is the process through which the zygote—a single cell—undergoes cell division without separation of the daughter cells to give a multicellular cluster of cells, followed by successive differentiation of the daughter cells into multiple cell types. This process is reminiscent of the process described earlier for the transition of single-celled protist into a multicellular organism. Transient multicellular colonies do not always have gametes and do not usually reproduce sexually.

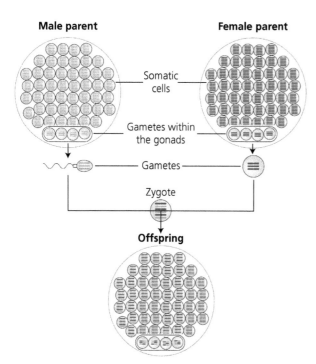

Figure 5.2 Sexual reproduction. In sexually reproducing organisms, two parents, each a multicellular organism composed of genetically identical cells, contribute a single gamete to the offspring. The gametes contain only half of the genetic composition of the parent cells and are formed in specialized structures—the gonads. Two gametes from two parents, normally a male and a female, fuse to give a zygote, which then develops to give an offspring, composed of genetically identical cells, each including genetic material from both parents.

Nonetheless, some form of meiosis or gametogenesis, and some form of exchange of genetic material between individuals is found in many groups of protists, often with conserved mechanistic aspects. This makes it highly likely that the evolutionary origins of sexual reproduction are earlier than the evolution of multicellularity itself.

Is there an advantage to sexual reproduction?

A well-known evolutionary conundrum is the selective driving force behind sexual reproduction. Organisms with sexual reproduction are more common and more diverse because the mixing of genetic material generates variability, and variability is the raw material of natural selection, thus accelerating the rate of evolution. While this explains the success of sexual reproduction, it does not explain how it first evolved and what advantage it conferred when it did.

From the cell's perspective, the fact that most cells in the body sacrifice their ability to pass on their genetic material in favor of a small number of cells

is surprising. This could only be viable from an evolutionary point of view if all cells in the organism are genetically identical. This is in fact the case in clonal organisms (which all complex multicellular taxa are). When a cell sacrifices reproduction in favor of supporting the entire organism's chance of reproducing, it is passing on its own genetic material via the genetically identical reproductive cells in the gonad.

From the individual organism's perspective, things are more complicated. Every gamete has only half of the individual's genetic material, and the offspring produced sexually will have half of its genetic material contributed by another individual, which is essentially a complete stranger. This sacrifice of half of the individual's genetic material is difficult to explain. Over the years there have been numerous theoretical explanations, often supported by mathematical models, which aim to clarify what makes this sacrifice selectively advantageous in the short term. We will not go into the details of these explanations here, but suffice it to say that there is still no conclusive and universally accepted answer to the question of what is the advantage to sexual reproduction.

What does a multicellular animal need?

With their large body size and diversity of cell types, multicellular organisms can carry out many more functions than their single-celled ancestors. But with this increased ability come increased needs. We started Chapter 3 by pointing out that the basic needs of a biological organism are fairly minimal—a source of energy and simple chemical building blocks. Multicellular organisms have to allocate cells and tissues to more complex organ systems that will allow them to obtain and utilize these basic needs more efficiently. We will briefly go through the requirements of complex organisms, focusing here (and for the remainder of the book) on animals only. Each one of these systems will be elaborated upon in subsequent chapters of the book.

Larger organisms deal with larger food and therefore require a compound system for acquiring, breaking down, and digesting the food. This is done by the **digestive system**, which we here take to include all the functions of food capturing and processing. An important ingredient in metabolic processes is oxygen, so that in parallel with acquiring food, animals must acquire oxygen. The main product of metabolic processes is carbon dioxide, which must be removed. Oxygen intake and carbon dioxide removal are done by the **gas exchange system** (or **respiratory system**). Organisms that are above a certain size cannot rely on diffusion for the distribution of the products of digestion or of essential oxygen to the entire body, and therefore make use of a **transport system**, or **circulatory system** to deliver materials to the entire body. To find food, and more generally to obtain information about the environment, animals make use of a **sensory system**, which usually allows them to collect several different types of information. Once food is located, the organism usually has to make its way toward the food source using its **locomotory system**. Integrating data from the sensory system and coordinating the resultant locomotory activity is done by the **nervous system**. Metabolic processes often result in waste molecules that need to be removed. This is the role of the **excretory system**. In a complex and competitive biological world, an organism has to protect itself from external threats and reduce the chance of it becoming the prey of another organism. Many of these needs are served by the body covering or **integument**. Finally, meeting all these needs is pointless if the organism does not ultimately pass on its genetic material to the next generation. This is the role of the **reproductive system**, which we have already touched upon in this chapter, and will come back to separately for each group of organisms we discuss.

Box 5.1 Choanoflagellates

Choanoflagellata is a taxon of unicellular protists that has been a focus of interest for over a century because of its supposed relationship to multicellular animals. When they were first discovered in the mid nineteenth century, scientists noted that their structure—consisting of a "collar" of hair-like protrusions known as **microvilli**, and a single long flagellum—is reminiscent of a type of cells known from sponges, the choanocyte (see Chapter 6). This similarity led to the suggestion that they may be the closest living unicellular relatives of Metazoa, and as such, the living organisms that are potentially most similar to the metazoan ancestor. It took over a century for this suggestion to be confirmed by molecular phylogenetics, and it is now broadly accepted. As the sister group to animals, choanoflagellates have the potential to inform us about the earliest stages in the evolution of multicellularity. We would expect them to have most of the necessary minimal toolkit for being multicellular, with any elements that they don't possess being metazoan novelties.

Indeed, much of what we know about the evolution of multicellularity in animals (some of which is covered in this chapter) comes from studies on choanoflagellates. Some choanoflagellates are able to form colonies, which are normally temporary. These colonies form by cell division and not by aggregation, just as expected from the accepted model of the evolution of multicellularity. Their genomes encode cell adhesion molecules that are also found in metazoans, and these are presumably involved in the formation and maintenance of colonies. Some species of choanoflagellates can also undergo a transition between several different cell states, again consistent with the model for the evolution of multicellularity presented above. Research into the biology of Choanoflagellata is sure to continue to provide insights into the early evolution of animals.

CHAPTER 6

Porifera

The simplest multicellular organisms

Overview

The sponges comprise the phylum Porifera. They are multicellular animals (metazoans) with a relatively simple body plan and organization. Sponges have significantly fewer cell types than most other animals, and these cells are not arranged into complex organs as in almost all other metazoans. Their body structure is not fixed, so there is either no symmetry at all, or there is radial symmetry. They are often presented as the simplest possible animals (but see Box 6.1), and some researchers believe they provide an example for what the earliest metazoans may have looked like. However, sponges are so different from all other animals that reconstructing the transition between a sponge-like early metazoan and any of the main metazoan groups is difficult. Despite their simplicity, sponges are a successful group of animals, with close to 10,000 described species in four generally accepted classes. They are all aquatic and almost exclusively marine, with members of only one taxon making the transition to the freshwater environment. All adult sponges are sessile and live attached to the substrate and, with very few exceptions, feed by filtering small organic particles and microorganisms out of the water that they actively pump through their porous bodies.

Sponge body organization

The structure of sponges is based on a body surrounded by a sheet-like layer of cells, known as an **epithelium** (see Chapter 7), and supported by a skeleton made up of numerous small biomineralized spicules, or of **spongin**—a unique, collagen-like protein—or of both. This skeleton is formed within a collagenous extracellular matrix, known as the **mesohyl**. The outer surface of the sponge is perforated by numerous pores, known as **ostia** (singular **ostium**), through which water enters the sponge's body. These pores are the source of the phylum's name, Porifera being Latin for "pore bearing." Ostia lead into a collecting space called the **subdermal cavity** and from there into a series of canals to small chambers where pumping of water takes place. Water is extruded from the chambers via other canals to a single main opening known as the **osculum** (plural **oscula**). As water flows through the cavities of the sponge body, food particles are filtered out and digested.

Many sponges are roughly vase or tube-shaped, and a single individual may be composed of several vases or tubes interconnected at their bases or be formed of branched tubular structures. Some sponges have a flattish structure and encrust the surfaces of rocks or dead corals. Others have an irregular shape with no clear axes—a shape known as massive.

Sponges are often classified into three types based on the arrangement of their chambers (see Figure 6.1). Note that this is not a taxonomic division, and these types are actually found in only one of the sponge classes, Calcarea. The simplest form is the **asconoid** type (or simply **ascon** sponges). In this type, the pores lead directly into the animal's central cavity, and there is a single osculum per cavity. Most asconoid sponges are highly branched, such that a single individual consists of multiple single tubes with many oscula.

Organismic Animal Biology. Ariel D. Chipman, Oxford University Press. © Ariel D. Chipman (2024). DOI: 10.1093/oso/9780192893581.003.0006

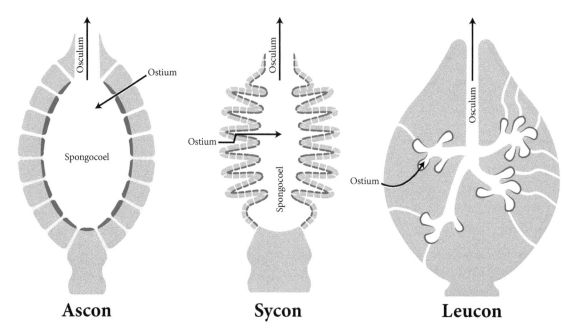

Ascon   **Sycon**   **Leucon**

Figure 6.1 Sponge body organization: The main openings and cavities of sponges and how they vary in the different types of organization. The red layer is the choanoderm and the blue layer is the pinacoderm.

In another type, known as **syconoid**, the ostia open into long, finger-shaped chambers and the chambers lie in a radial pattern surrounding a central cavity. Water flow is from the outside into the chambers, where food particles are filtered out of the water; the water then continues to the central cavity, or **spongocoel**, and out through a single osculum.

The **leuconoid** or **leucon** arrangement is the most intricate type of canal system in sponges and is found in some of those that are most commonly encountered and in many of the largest sponges. Leuconoid sponges have canals that reach from the sponge surface to all regions of a thickened body wall, some short and others long. Incoming water passes through canals to a single chamber and on through excurrent canals before reaching the central cavity and being vented from the osculum. In some leuconoid sponges, the central cavity or spongocoel is highly reduced to lie just under the osculum. Often, multiple oscula, and multiple canal systems, share a common outer epithelium (**ectosome**), which allows these sponges to become quite large.

Cell types in sponges

There are five main categories of cell type in sponges, although within each of these categories, cells are diversified into a range of more specific variants (Figure 6.2). The outer and inner surfaces of the sponge are made up of epithelial cells, known as **pinacocytes**. The pinacocyte layer is known as the **pinacoderm**. Pinacocytes can be further classified based on the specific surfaces they form.

The inner layer of the main cavity, of most other chambers and of some of the canals connecting these chambers are lined by cells known as **choanocytes**. These have a flagellum and a collar of microvilli. The beating of the choanocyte flagella is the driving force for the movement of water through the sponge. The microvilli of the choanocytes are covered by mucus that traps food particles flowing through the choanocyte-lined chambers. Thus, these cells are responsible for the two main feeding related functions in the sponge: delivering food into the body and trapping it there. The layer of choanocytes is known as the **choanoderm**.

The incurrent pores or ostia in ascon sponges are made up of unique cells known as **porocytes**.

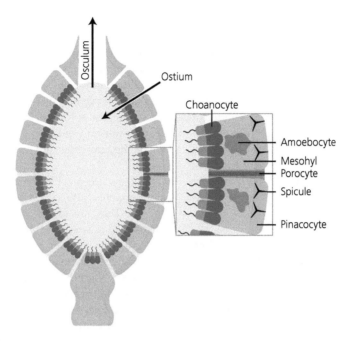

Figure 6.2 The main cell types in sponges.

These cells form early in development as flat cells, which roll up to make tubes. The porocytes form pores through the pinacoderm, the choanoderm, and the intermediate acellular mesohyl. In sponges with more complex organization, the ostia are made up of multiple cells.

Within the collagenous mesohyl, there are motile cells of several types. The most common are the **amoebocytes** (also known as **archeocytes**), which are responsible for the digestion of food particles caught by the choanocytes. There is no extracellular digestion in sponges. The amoebocytes can differentiate into several different cell types, including choanocytes and pinacocytes, and these cell types can dedifferentiate to become amoebocytes.

Also important within the mesohyl are the cells that secrete the spicules that form the skeleton. Different sponge taxa are differentiated by the material used to form the skeletal spicules. The spicules can be either calcareous (calcium carbonate based) or siliceous (silicon dioxide based). In addition to the mineral spicules, many sponges have an organic skeleton made up of spongin. The specific shape and organization of the spicules are very distinctive and are often used as characters for defining and identifying sponges at the family, or even species level.

An interesting characteristic of sponges is that upon dissociation, cells can reaggregate to form a functional sponge, with differentiated cell types. This behavior highlights the plasticity of sponge cells and the ease with which they can change form and function, relative to most other metazoan taxa, where cell identity is much less labile.

Life history and ecology of sponges

The majority of sponges are sessile suspension feeders. They usually attach permanently to hard substrates and do not leave them. However, the active movement of water through their porous bodies, driven by the constant beating of thousands of choanocyte flagella, makes them extremely efficient suspension feeders, able to process vast amounts of seawater relative to the animal's size. Estimates suggest that the volume of water that passes through any individual chamber in a sponge in a single day is over 1000 times the volume of the chamber (some estimates put the number at 50,000 times the volume). Food particles suspended in the water, including organic detritus, bacteria, or larger single-celled organisms, are trapped by the choanocytes and digested through **phagocytosis** either by the choanocytes themselves or by amoebocytes moving through the mesohyl.

Most sponges grow slowly without clear axes of growth, although some have been reported to develop rapidly and overgrow neighboring organisms. Some sponges are annual and seasonal, whereas others may live for a few decades. Some of the largest sponges in the world, the Caribbean barrel sponges, are estimated to be a few centuries old.

Sponges can reproduce sexually and asexually. Asexual reproduction is either through budding, where a group of cells grow and pinch off the main body of the sponge, or through pieces of the animal simply breaking off and then rearranging and redifferentiating to give a new functional sponge. This asexual reproduction is presumably the basis for the cellular plasticity of sponges described above.

Most freshwater sponges are able to generate small structures known as gemmules, which are dormant reproductive structures that can survive desiccation and freezing. Once suitable conditions return, the gemmules hatch and undergo differentiation to give rise to a small sponge.

In sexual reproduction, specific cell types differentiate to form gametes. There are no distinct gonads or reproductive organs in sponges. Normally, choanocytes differentiate and undergo meiosis to form sperm, and amoebocytes undergo meiosis to form ova. Some sponge species are **hermaphrodites**, meaning the same individual can generate both sperm and ova, though this is usually sequential and not simultaneous. In other species, males and females are separate individuals, and there is no transition between the sexes. In some species, sperm are released into the surrounding water through the osculum and are taken up by other individuals of the same species through their ostia. The sperm cells are caught and encapsulated by choanocytes, which deliver them into the mesohyl where they find and fertilize the ova. The fertilized zygotes are then released into the water. In other species, both sperm and ova are released into the water. Fertilization in such species is external and does not involve the participation of other cells.

The fertilized zygote undergoes several rounds of cell division and minimal embryonic development to form a simple free-swimming ciliated juvenile or **larva** (plural **larvae**). Different groups of sponges have different typical larval forms, but in all cases, they are bilayered ciliated larvae that are either flat or roughly spherical. The larvae are carried by the currents and settle on suitable hard substrates, where they can grow and differentiate into an adult sponge.

Sponges are most common in shallow marine environments although there are also deep-sea species. They are found in most marine environments in all oceans, from the tropics to polar regions, and are often a major component of the environments where they live.

Sponges form symbiotic relationship with a range of single-celled organisms. Frequently, these endosymbionts are involved in primary production. Sponge endosymbionts may form large and complex microbial communities, with a biomass that sometimes nears the biomass of the host sponge. The details and mutual benefits of the relationships between sponge and symbionts are not well understood.

Relationships between a sponge and its neighboring sponges, whether of the same or different species, can be of an antagonistic nature, with each sponge secreting toxic compounds in order to limit the growth of its neighbors. Indeed, sponges can produce a remarkable range of chemicals, either as competitive agents or as toxic compounds, to protect themselves from being eaten. Sponges are among the most biochemically complex organisms known, and they are seen as a treasure trove of natural compounds, offering a potential for the discovery of various beneficial compounds for medicine or industry.

Possibly related to their biochemical complexity, sponges are also among the most colorful organisms on the seafloor. They can be found in almost every color imaginable, and sponge communities are often a dazzling mix of bright colors. The colors stem either from various pigments within the mesohyl or from colorful endosymbiotic algae. These bright colors might have been selectively favored to serve as a warning to potential predators, but they could simply be inevitable (non-selected) characteristics of some of the sponges' numerous bioactive compounds.

Sponge diversity and taxonomy

Traditionally, sponges have been divided into three classes, but most recent analyses provide support for four distinct monophyletic classes (Figure 6.3).

Demospongiae is the largest and most diverse group of sponges. Their skeletons are usually made of siliceous or organic spicules, or a combination of the two. These spicules can appear together with (or within) fibrous spongin. Sometimes, there is an integral skeleton (not separate spicules) made of calcium carbonate, and in some cases, there is no skeleton. The familiar bath sponges are members of this group. Demospongiae also includes the only sponges to have made the transition to fresh water. Freshwater sponges have undergone significant physiological adaptations to the challenging salt-poor environment. They are composed of a layer of cells covering the mesohyl, and normally encrust surfaces or suitable objects, rather than having a typical vase shape. Rarely, freshwater sponges form larger branched structures.

The glass sponges or Hexactinellida are the most diverged sponges in terms of their organization. Their siliceous skeleton is made up of typical six-pointed spicules (hence the name). The entire animal is composed of **syncytial** structures (tissues with no membranes between adjacent cells). Some of the typical sponge cell types can be found as cells without a nucleus, but other than that, they diverge significantly from most common sponge characteristics. Hexactinellids are found almost exclusively in the deep sea and flourish in temperate to polar regions. Their diversity is low relative to other sponge classes, with only a few hundred known species.

Calcarea, as the name suggests, includes sponges with calcareous (calcium carbonate based) skeletons. Members of this group are found mostly in shallow waters. All three types of body organization are found in this group.

The most recently identified class of sponges Homoscleromorpha was previously characterized as a subclass of Demospongiae. However, both molecular phylogenies and morphological and embryological comparisons place them as a distinct group. Homoscleromorph sponges have unique siliceous spicules that are structurally distinct from those of demosponges. This is a very small class with fewer than 100 species described.

There is almost full consensus about the monophyly of sponges. A clear majority of molecular phylogenies support Porifera as monophyletic, with

(a)   (b)

Figure 6.3 Examples of sponge diversity: (a) A demosponge, the yellow tube sponge *Aplysina fistularis*; (b) A homoscleromorph, *Corticium candelabrum*.

Source: a: Photo supplied from Shutterstock: scubaluna (https://www.shutterstock.com/image-photo/tube-sponge-aplysina-fustularis-125223833); b: Photo supplied from Shutterstock: Ana y Erik (https://www.shutterstock.com/image-photo/close-view-orange-sea-sponge-macro-1994716829).

Calcarea placed as a sister group to Homosclero-morpha, and Hexactinellida as sister to Demospon-giae. Some molecular analyses have suggested that sponges are paraphyletic, but this idea is not supported by more recent and thorough analyses.

Evolutionary history

Sponges are probably the multicellular animals with the oldest fossil record. Their record begins with trace molecules or biomarkers, of a type of organic compounds known as steranes believed to be unique to demosponges. These biomarkers have been found in rocks from the Cryogenian, over 635 million years ago (MYA). This date is compatible with molecular clock estimates for the origin of sponges. Putative sponge body fossils are found in the Doushantuo deposits of China from the early Ediacaran (~575 MYA). The earliest reliable sponge spicules are found from the early Cambrian, about 530 MYA, onward. Both calcareous and siliceous sponges (including glass sponges) have a long and diverse fossil record following the Cambrian. A group of sponges with hard calcareous skeletons, known as stromatoporids, were the first reef-builders in the Paleozoic seas, reaching their peak in the Silurian (440–420 MYA). Sponge spicules are very common in the marine fossil record of all post-Cambrian periods and can be used as stratigraphic markers.

Sponges are generally believed to be the sister group to all other metazoans, and as such to represent a possible model for the early evolution of multicellularity. The similarity of sponge choanocytes to choanoflagellates (see box 5.1) and the recent discovery that choanoflagellates can differentiate into several types of cells, including amoeboid cells, allow us to attempt to reconstruct the stages through which sponges, and multicellular animals in general, evolved. However, the view of sponges as the most primitive metazoans has been challenged recently on phylogenetic grounds (see Box 9.1), raising the intriguing possibility that sponges may have actually undergone secondary simplification very early in their evolutionary history. Either way, it is clear that sponges hold a key position for our understanding of early animal evolution.

Box 6.1 Placozoa

Placozoans are an enigmatic group of multicellular animals, with an extremely simple body organization. Placozoans are, as their name suggests, flat animals. They are disk shaped and are made up of two main layers of epithelial cells with a small number of cells between the two layers. Until recently, the phylum Placozoa was known to include only a single species, *Trichoplax adherens*. It is now suspected that there are many more morphologically similar species, as well as a small number of somewhat more complex species that have variously branched bodies.

The two epithelial cell layers are ciliated. The cilia of the lower (basal) layer form the primary locomotory structure, and the animals move across the substrate through ciliary gliding. The lower cells also secrete digestive enzymes and can invaginate partially to serve as an ad-hoc digestive system. In between the two layers, there is a simple mesenchyme layer. Placozoans have six identified cell types.

There is no defined symmetry, no anterior or posterior, no tissues, and no organized system of any kind. Placozoans reach a few millimeters in diameter and are made up of a few thousand cells at most.

Almost all phylogenetic analyses place placozoans strictly within Metazoa, closer to all other metazoans than sponges. This indicates that placozoans have most likely descended from more complex animals and have undergone extreme secondary simplification. Placozoans also have very simple genomes, although there is evidence to suggest that these are primitively simple and secondarily simplified.

A surprising hypothesis about placozoan origins comes from an analysis of the Precambrian fossil *Dickinsonia*. This fossil belongs to a group of organisms referred to as the Ediacaran biota and dates to about 550 million years ago. The fossils of the Ediacaran biota mostly cannot be assigned to any modern taxa. However, *Dickinsonia* may be one of the earliest bilaterians or a sister group to them. Analysis of trace fossils—imprints made by the movement of *Dickinsonia* across the substrate—suggest that it fed like placozoans through secretion of digestive enzymes to break down microorganisms on the substrate. While the resulting hypothesis requires a large number of unsupported assumptions, it is possible that placozoans evolved through the simplification of an ancient primitive *Dickinsonia*-like bilaterian.

CHAPTER 7

Germ layers

Inside and out

Germ Layer Theory

The idea that the embryo—and the adult animal—can be divided into a series of distinct cell populations dates back to the early days of comparative embryology in the mid nineteenth century. According to this idea, the **germ layer theory**, at an early stage in embryonic development, the embryonic cells undergo differentiation into three germ layers. These germ layers maintain their identity throughout the developmental process and give rise to a series of layer-specific tissues. The **ectoderm** gives rise to tissues that will be located on the outside surface of the embryo, and later to the animal's external covering, the integument, as well as the nervous system and several other structures. The **endoderm** gives rise to tissues that will define the developing digestive system. Later, these cells will form most of the inner lining of the digestive system and contribute to many digestive organs. The **mesoderm** is responsible for the majority of cell types and tissues. It will form most of the intermediate tissue, the muscles, and many internal support structures.

We will start with the first two mentioned germ layers, the ectoderm and endoderm, which form the inside and the outside of the organism. These layers have some characteristics in common, and they are the only germ layers found in the first organisms we discuss. After introducing them, we will elaborate on the integument and the digestive system more generally.

Epithelia

The main similarity between the ectoderm and endoderm is that they are composed of **epithelial** cells and form tissues known as **epithelia** (epithelium in the singular). This is true both of the early embryonic ectodermal and endodermal cells and of most (but not all) adult structures they form. Epithelia are two-dimensional sheets of cells and are often found lining cavities or forming borders between different organs. Epithelial cells are polar cells, meaning they have a distinct direction. Every cell has a **basal** pole and an **apical** pole. The basal pole is usually attached to a flat acellular matrix, known as the **basement membrane**, which acts as an anchoring surface for the entire epithelium. In simple epithelia, the apical pole faces and interacts with the environment. This can be either the external environment or an internal environment such as the gut lumen. The apical pole is the part of the cell that is involved in secretion or absorption, in sensing the environment and in protecting from it.

Epithelia can be single layered or multi layered. In a simple, single layered epithelium, the basal pole of all the cells rests on the basement membrane and the apical pole of all the cells faces towards a cavity or the external environment. In multilayered epithelia, only the inner layer rests on the basement membrane and only the outer layer faces the environment, while other cells lie with basal and apical poles of adjacent cells touching. Even in multilayered epithelia, the tissue is no more than a few cells deep, and the structure is still mostly that of a two-dimensional sheet.

Epithelial cells can be further divided into types based on their relative dimensions. **Squamosal** cells (literally, scale-cells) are wider than they are tall and often form the outer layer of a multilayer epithelium facing the outside world. **Cuboidal** cells are

Organismic Animal Biology. Ariel D. Chipman, Oxford University Press. © Ariel D. Chipman (2024). DOI: 10.1093/oso/9780192893581.003.0007

roughly the same width and height. **Columnar** cells are taller than they are wide and are often involved in secretion.

Epithelia are fairly diverse, but they make up only a small part of the entire diversity of animal tissue types. We will leave discussion of non-epithelial cells and tissues for later in the book. Meanwhile, for the sake of simplicity, we will refer to all non-epithelial tissue as "connective tissue," although there is much more to it.

As we said, epithelia often line cavities and form borders. The borders can either be between different sections of the organism's body or between the body and the outside world. As we shall soon see, the outside world can also be an internal cavity that is connected to the outside, such as the digestive tract. In fact, if we were to describe an animal in the most basic topological terms, we could say that it forms a tube. The inside of the tube is the digestive system, lined by endodermal cells, and the outside of the tube is the integument, with ectodermal cells forming the border with the outside world. We will now discuss these two main organ systems, both of which are based on epithelia, but with contributions from other tissues as well.

The outer border—integument

The outer covering of the organism is known as the **integument**. The integument as an organ system contains more than just the external epithelium. It often includes supportive connective tissues, acellular components, and dedicated non-epithelial cells embedded within the epithelium. The integument has several important roles. The first is to form a physical boundary that separates the organism from its environment and provides mechanical protection from the outside world. It usually also acts as a selective chemical boundary, regulating which substances enter and exit the body.

In many cases, the integument houses or forms sensory organs. These are cells or groups of cells that collect information about the environment through a range of different channels. We will discuss these channels and the cell types that mediate them in Chapter 12. In addition to collecting information from the environment, the integument provides information to other organisms, communicating

with members of either of the same species or of other species. This can be through visual communication, as the integument is the part of the organism visible to other organisms. The integument often includes **pigmentation**—colored substances that can be seen by others. These colors can serve to hide the organism through camouflage, indicate that it is dangerous and should be left alone, or signal attractiveness to potential mates. The integument can also produce chemical signals—diffusible molecules that are picked up by other organisms and communicate information that can attract or deter other individuals. In some cases, the integument houses sound producing structures, which again provide information to other individuals.

A less common role, but one that is familiar to us as mammals that maintain a constant body temperature, is **thermoregulation**. In mammals and in birds, the integument produces structures that provide an insulating layer against heat loss—hair and feathers, respectively. In addition, the mammalian integument produces sweat that helps cool the animal off.

Many animal integuments include secretory cells—cells that synthesize and secrete specific chemicals. Clusters of cells that secrete together are known as **glands**. These cells or glands can have a role in many of the functions described above: chemical communication, protection, or thermoregulation.

To help understand how all these functions interact in a single tissue, we will elaborate on two examples of fairly complex integuments: the integuments of mammals and of insects. In mammals (Figure 7.1), the outer portion of the integument is a multilayered epithelium knows as the **epidermis**. This is the part of the integument that is derived from the embryonic ectoderm. The epidermis lies on a tough and elastic basement membrane. The deep cells are cuboidal and divide constantly. As they divide, the upper cells are pushed up and gradually become more squamosal in shape and accumulate a fibrous protein known as **keratin**. The keratin forms a waterproof external layer that in many cases is also very thick and provides physical protection. The outer keratinous layer is sloughed periodically and replaced by new cells rising from the basal parts of the epithelium. Below the epidermis is a thick

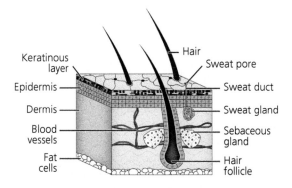

Figure 7.1 The structure of mammalian integument: A schematic three-dimensional section of the main components in the integument of mammals.

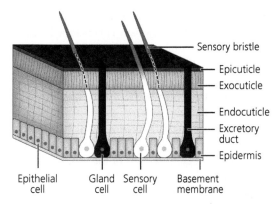

Figure 7.2 The structure of insect integument: A schematic three-dimensional section of the main components in the integument of insects.

layer of connective tissue knows as the **dermis**. It is in turn divided into an upper and lower layer, each with a different ratio of connective tissues and elastic fibers. In the epidermis are deep pockets—hair follicles—that sink into the dermis and secrete keratin in a structured manner to form hair—a unique mammalian structure. Beside the hair follicles are secretory glands—sebaceous glands—that produce an oily substance that coats the hairs and makes them water resistant. Pigment cells of various types provide color either to the hair itself or to the surrounding dermis. Also in the dermis are sweat glands that secrete salt rich water via ducts that open to the surface of the epidermis. Finally, the dermis includes a variety of touch receptors, heat receptors, and others that provide tactile information about the near environment. This description is naturally very general. Based upon this general structure, we can find numerous variations, giving us the diversity of mammal skin and fur types.

The insect integument (Figure 7.2) is very different from mammalian skin, despite including many of the same components. At its base is a single-layered epidermis that sits on a thin basement membrane. The epithelial cells are columnar secretory cells whose main role is to secrete the overlying cuticle which forms the bulk of the insect integument. The cuticle is a hard and inflexible multi-layered, acellular structure, composed mostly of a complex carbohydrate called **chitin**. In addition to the chitin-producing cells, the epidermis includes other types of secretory cells that produce a range of cuticular secretions, some having a role in chemical signaling

and others coating the cuticle with a waterproof waxy substance. Embedded in the epidermis are sensory receptor cells that are connected to external hairs of different types. Note that while in mammals, the hairs are made of keratin and serve mostly for thermoregulation (and to a lesser extent for tactile sensing), in insects they are made of chitin and serve almost exclusively for tactile and chemical reception. The insect cuticle is molted periodically. This is done through the secretion of compounds that dissolve the deep layer of the cuticle, thus separating it from the epidermis. The outer layers of the cuticle are shed, and a new cuticle is produced by the epidermis.

The inner border—digestive systems

Although we tend to think of the digestive system as being "inside," it is actually external to the organism, since it connects to the outside world with one or two openings—the mouth and the anus. The digestive system has both secretory roles and absorption roles. It secretes enzymes and acid to digest the food, and then absorbs the digested products. In most cases, there is a spatial division between the secretory (digestive) role and the absorptive (nutritive) roles.

We will start with a cross section through the tissues that make up the digestive system (Figure 7.3). Most regions will have at least some of these tissues, and many will have all of them. The outermost layer of the digestive system is the layer that is in contact with the digestive cavity—the outside world as far as the organism is concerned. It is a single-cell-thick columnar epithelium derived from the endoderm. Depending on the region, these cells can either be devoted to secreting or to absorbing. Unlike the outer layer of the integument, it is usually relatively soft and elastic and does not provide a physical barrier. However, it is often covered by a mucous layer, which protects the cells from the harsh environment of the gut, while also acting as a selective chemical barrier. Below this epithelium is a layer of connective tissue of varying thickness. This connective tissue may include multi-cellular secretory glands. Beyond the connective tissue is a **muscle** layer, usually made up of a layer of longitudinal muscles which contract along the axis of the digestive tract and a layer of circular muscles which contract to narrow the digestive tract. The combined action of these two muscle layers pushes the food along the gut from the mouth to the anus.

The digestive system is adapted to the specific nutrition of the organism; what the animal eats, how it obtains it, and what level of processing the food needs before digestion. It can be divided into a number of conserved regions, which are found to a greater or lesser degree in different animals. While these regions are not homologous across different organisms, and they can vary significantly in detail, the general layout of the digestive system is roughly similar. Rather than giving specific examples, in what follows, we will provide a general overview of the different elements that are found in the digestive system of most animals (Figure 7.4).

The first role of the digestive system is to acquire and ingest food. This is done by the **mouth** and its accessory structures. Often these are responsible for the capture of food, for holding on to it, and to pushing it into the digestive tract. The ingested food is brought into the **pharynx**, the first distinct region of the digestive system. The pharynx is where the food undergoes preliminary processing. It can be broken down into manageable pieces, or undergo preliminary chemical breakdown. The pharynx can be distinguished from the rest of the digestive system by the fact that it is usually derived from the ectoderm, and often includes elements that are distinctively ectodermal. From the pharynx, the food is transferred deeper into the system, via a section known as the **esophagus**. Usually, the esophagus has no specific role, and is merely an intermediate stage between preliminary processing and the actual digestion. However, in some cases, it takes on the role of mechanical or chemical processing, and the food is stored there for a certain amount of time before moving on. In those cases, it is known as the **crop**.

The **stomach** is the region where more intensive digestion takes place. The lumen of the stomach is usually extremely acidic, and the acid is

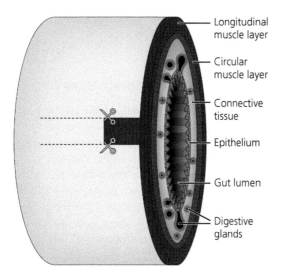

Figure 7.3 Schematic section through a generalized digestive system: This section is not meant to represent a specific region of the digestive system, but includes elements found in most parts of most digestive systems.

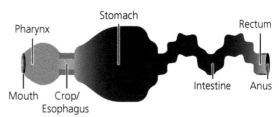

Figure 7.4 The main regions of a typical digestive system: This highly generalized structure of a typical digestive system, indicates the division into main regions and the common elements included in the digestive systems in many animals.

supplemented by a diversity of enzymes, dedicated to breaking down specific macromolecules into smaller usable molecules. The epithelium of the stomach is devoted almost entirely to secretion. The stomach lumen tends to be wider than any other region of the digestive system, and the food often spends much time sitting in the stomach being digested. Digestion is normally aided by endosymbiotic organisms, either bacteria or single-celled eukaryotes, in the stomach or in the crop. The symbionts help in breaking down compounds that cannot be broken down by the animal's own enzymes. These are often complex plant compounds such as cellulose or lignin. Because of the durability of these plant compounds, their digestion sometimes includes a complex cycle of processing, digestion, and re-processing, as seen in ruminant mammals.

Once the food has been broken down in the stomach, it moves on to the **intestine**, whose main role is to absorb the digestive products in preparation for the transfer to the rest of the body. The lumen of the intestine is not as acidic as that of the stomach, and the epithelium lining it is dedicated to absorption. The distinction between digestion and absorption is not always as sharp as presented here, and there can be absorption in the stomach and digestion in the intestine, but this type of regionalization is the case in most animals. In order to aid in absorption, the surface area of the intestinal epithelium is increased. This can be done through complex infolding of the intestinal wall to give a convoluted surface, through twisting of the intestines inside the body to increase its length, or through a range of other strategies. In addition, the inner surface of the intestine is covered by finger-like protrusions at the cellular and sub-cellular level, known as **villi** and **microvilli** respectively. The final section of the intestine is sometimes specifically devoted to reabsorbing water from the digested mass, in order to avoid too much water loss through the digestive process. This section, when present, is known as the **rectum**.

All undigested products are removed from the body through the posterior opening of the digestive system, the **anus**. The undigested products are known as **feces**, and they can have many different shapes, colors, textures (and indeed, smells), based on the nature of the food and the structure of the digestive system. Note that this disposal of digestive waste is distinct from the excretory system, which deals with removing toxic waste products from the body itself. We will discuss this system in Chapter 28.

There are numerous variations on this general description. Animals that eat difficult to digest substances (e.g., termites that eat wood or goats that eat grass) have expanded stomachs and more of the digestive process is devoted to breakdown of the food. Animals that eat nutrient-poor food (e.g., earthworms that extract organic substances from soil) will have longer intestines to absorb as much as possible from the food as it passes through the digestive tract. Some organisms have an undifferentiated digestive system, where both digestion and absorption happen in the same space. This is often the case in animals with a single opening to the digestive system (the cnidarians we will meet in Chapter 9). A system with only one opening is probably the primitive state for animals, but it has re-evolved secondarily in several animals whose ancestors had a through-gut.

Modes of feeding

Animals can be classified based on the type of food they eat and the way they obtain it. These are very broad classifications, and like any broad classification, we can find examples that are intermediate between different classes or that don't neatly fit into any of them. Nonetheless, as we will see throughout the book, animals with similar feeding modes have convergently evolved similar feeding structures. Looking at an animal's digestive system, and especially at the mouth and accessory structures, is usually enough to tell us what an animal eats.

Many aquatic animals collect food particles that are suspended in the surrounding water. These can be minute planktonic animals or plants, single-celled organisms, or pieces of organic matter. We refer to this type of feeding as **suspension feeding** or **filter feeding**. Animals with this feeding mode usually have an accessory structure such as net or a series of arms or **tentacles** near the mouth that collect particles and bring them to mouth for ingesting.

Eating prey that is larger than suspended particles is usually referred to generally as **macrovory**. This is often divided into **carnivory**—eating other animals—and **herbivory**—eating plants. The mouth and pharynx will normally be adapted to handling large food items, with teeth, a beak, or some similar structure. Plant material is more difficult to digest, therefore herbivores will usually have longer digestive systems, with specialized regions for mechanical and enzymatic breakdown of the food.

In aquatic environments and in rich and damp terrestrial environments, the substrate is often covered with partially degraded organic material—detritus. Feeding on this organic material is known as **detritivory**. Detritivorous animals usually have relatively simple mouths, since the food requires little effort to acquire and is already partially degraded. Many detritivorous animals live within the substrate, earthworms being a familiar example.

CHAPTER 8

Motility and symmetry

Why move?

Most animals move, at least to a certain extent. Movement is sometimes listed (albeit erroneously) as one of the characters that differentiate animals from plants and other immobile organisms. We should however be careful to differentiate between **movement**, which can be any change in the position of parts of an organism relative to others, and **locomotion**, which is the autonomous movement of the entire organism from one place to another. Plants move (e.g., trees swaying in the wind). Seeds and spores disperse with the wind. Only animals locomote, but not all of them do. The ability to locomote is known as **motility**, and it is different levels of motility and their consequences that concern us in this chapter.

There are several reasons why an animal should need to move to a different location. It can move to find food or other resources (foraging). It can improve its situation in general by moving to a more suitable spot. Animals move around in order to find mates or in order to disperse after reproduction. Often an animal will not move toward a new place but away from where it is, in order to escape predation or avoid toxic substances.

Types of motility

We distinguish between different levels of motility (Figure 8.1). Animals that do not move from their place at all are known as **sessile** animals. These include animals that are attached to the substrate such as sponges (Chapter 6) and corals (Chapter 9), or animals that spend their entire time inside a fixed chamber like a clam in a shell (Chapter 15), or a tube-worm in a tube (Chapter 17). Animals that are able

to move, but do so rarely or only for short distances are known as **sedentary** animals. Sedentary animals do not normally need to move large distances to forage for food and usually have some form of protection against predators, like the spines of a sea urchin (Chapter 26), so they don't need to escape to avoid being eaten. Animals that move freely in space are referred to as **mobile** or **motile**. Mobile animals are more diverse (and more familiar) than sedentary or sessile animals and are found in more types of environment.

Aquatic mobile animals can be divided based on where in the water column they live (Figure 8.1). We distinguish between **benthic** organisms, which spend most of their time on or in the bottom substrate, or close to it, and **pelagic** organisms which spend most of their time in the water column. Benthic organisms can be further divided into **epibenthic** organisms that live and move on the surface of the substrate, and **endobenthic** organisms that live beneath the surface and move by tunneling through the substrate. Pelagic organisms are divided into those that are **planktonic** and drift with the currents with little control of their own, and the more active **nektonic** animals that can swim freely with more control on their position and can actively move up and down the water column.

There are numerous terms used to describe locomotion in the terrestrial and aerial sphere (the world outside the water). "**Walking**" and "**running**" are two of several terms that refer to moving using limbs on land, with the differences between them having to do with the relative speed of movement. The border between walking and running is not sharply defined, and there are alternative terms for specific intermediate gaits (e.g., trotting, cantering, ambling,). **Crawling** is another poorly defined term

Figure 8.1 Types of motility: Different marine organisms demonstrating types of motility and position in the water column. (A) sedentary endobenthic worm. (B) mobile nektonic fish. (C) sessile endobenthic clam. (D) sedentary epibenthic sea urchin. (E) mobile epibenthic crab. (F) mobile planktonic jellyfish. (G) sessile epibenthic tube worm. (H) sessile epibenthic coral.

that is used to refer both to organisms walking on short limbs (as in insects) or to movement where the underside of the organism is in direct contact with the substrate (as in snakes). **Flying** is locomotion through the air using wings, whereas **soaring** is flying that doesn't involve active movement of the wings. We will expand on the mechanics of different types of locomotion in Chapter 16.

Locomotory systems

Animals use a number of different types of locomotory systems. The most primitive system is **ciliary motion**. In this type of locomotion, ciliated epithelial cells provide thrust through the synchronous movement of numerous cilia. This is similar to the type of locomotion we saw in some of the unicellular eukaryotes. Ciliary motion is mostly found in benthic organisms, where the surface facing toward the substrate is covered in cilia and the animal glides across the substrate. It is also found in some terrestrial crawlers such as some snails, where the animal secretes a mucus trail and uses its cilia to move along the mucus. It is found more rarely in planktonic organisms, where the cilia provide a limited amount of control over the direction of movement. No truly nektonic animals use cilia for locomotion.

Locomotion using **muscles** is far more common than ciliary motion. (We will expand on muscle-based locomotion in Chapter 16). Muscles are a distinct tissue type that is conserved and similar among many taxa. The cells that make up muscle tissue are known as **myocytes**. They are typically elongated and are filled with two fibrous proteins called **actin** and **myosin**. These two proteins are arranged in a partially overlapping linear configuration. Movement of actin over myosin requires energy and leads to the contraction of the cell. Muscle tissue is composed of many myocytes arranged linearly to create fibers. Synchronized contraction of the cells along

the muscle fiber, leads to a contraction of the whole muscle.

There are two main types of muscle: **smooth muscle** and **striated muscle**, although other types can be found and there are muscles that are intermediate between the two main types. In smooth muscle, individual cells are spindle shaped and are often arranged in sheets rather than in fibers. This type of muscle is found in the layers that line the digestive tract. Striated muscle (also known as reticulated muscle or skeletal muscle in vertebrates) is highly structured into hierarchically arranged fibers. The individual fibers are syncytial, meaning there are no membranes between the individual cells, and the fiber functions as a single multinucleated cell. Striated muscles are faster and usually more powerful and, because of their syncytial structure, are easier to control than smooth muscles. Striated muscles are therefore the main type of muscle used for locomotion in many animals.

Symmetry and its connection to modes of motility

The mode of motility has important consequences for the animal's body plan, most notably for the type of symmetry it displays. Sessile and sedentary animals often display **radial symmetry** (Figure 8.2). In an animal that does not normally move, all directions are equal. There is no front or back, and therefore no directionality. The outcome is an animal whose main axis is a rotational axis, with structures repeating radially around it. Even in sessile and sedentary animals that do not display radial symmetry, the feeding organs tend to be radially symmetrical, often in the form of a ring of tentacles surrounding the mouth. Radial symmetry is also found in some planktonic animals that drift with the currents and don't have a preferred direction

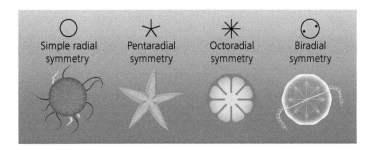

Figure 8.2 Different types of radial symmetry.

(jellyfish being a clear example). The single axis of radially symmetrical animals usually runs through a central mouth, which lies at one end of the rotational axis. The side with the mouth is known as the **oral** side, with the other side known as **aboral**. Consequently, the rotational axis is called the **oral–aboral axis**.

Radial symmetry is never perfect, unless the animal is an exact circle (and no animal is). There will tend to be a number of repeats around the rotational axis (Figure 8.2). Thus, we can have pentaradial symmetry (five repeats) or octoradial symmetry (eight repeats), for example. A special case of this is biradial symmetry, wherein an animal is radially symmetrical, but also has a plane of mirror symmetry running through the rotational axis. Even given the above distinctions, in the biological world, radially symmetrical animals are usually not truly radial. Very often, one direction will be somewhat different and break the perfect symmetry. The animal may have a preferred direction of movement or food capture, or there may be singular structures found on one side only.

Sponges, the sessile animals we met in Chapter 6, provide several exceptions to the principles we just detailed. Many sponges, have no symmetry whatsoever, and grow in a random and indeterminate matter. These sponges have no axes and no directionality. On the other hand, some sponges, such as barrel sponges, display almost perfect radial symmetry. Because there are no clear organs, nothing repeats around the rotational axis, and at the macroscopic level the sponge is truly radial.

In contrast with sessile and sedentary animals, mobile animals do have a clear direction, which requires a distinct front end and back end, and therefore an **anterior–posterior axis** (Figure 8.3).

Because animals move in a world that has gravity, and they are often affected by an underlying substrate, there is also a difference between the side of the animal that faces down, toward the substrate, and the side that faces up, away from it. The downward facing side of the animal is called the **ventral** side, and the opposite side is **dorsal**. In three-dimensional space, the definition of two axes forces the existence of a third axis. Since there are no inherent differences between the two sides of this axis, the left side and the right side, most animals with an anterior–posterior axis, tend to be bilaterally symmetrical. Note that in bipedal animals, such as humans, the ventral side of the animal will often face forward rather than downward.

A further difference between mobile and non-mobile animals lies in the structure of their nervous system. Sessile and sedentary animals tend to have non-centralized nervous systems. These animals will have a nervous system that is diffuse and arranged in a nerve net, with no main processing center. Mobile animals, with an anterior–posterior axis and bilateral symmetry, tend to have a nervous system that is concentrated along the midline of the main body axis, with an enlarged region in the anterior. We will discuss this point in more detail in Chapter 12.

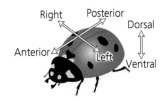

Figure 8.3 Body axes in bilaterian animals.

CHAPTER 9

Diploblastic organisms

Cnidaria and Ctenophora

Diploblastic organisms

Having introduced the basic concepts of germ layers in Chapter 7, we can now discuss in detail the animals that display only two layers: Cnidaria and Ctenophora. These animals are referred to as **diploblastic** animals or **diploblasts** (but note that this should not be treated as a taxonomic group). They have two embryonic layers of epithelial tissues, corresponding to the endoderm and the ectoderm, and all of the adult structures are derived from these two germ layers.

Diploblastic animals share a number of characteristics. All have radial symmetry of some sort, at least superficially, and all have a simple digestive system with only one main opening. The tissues of diploblasts are mostly gelatinous and they are therefore often referred to colloquially as "jellies." Despite these similarities, most phylogenies do not place Cnidaria and Ctenophora into a monophyletic group. Old taxonomic literature uses the terms Coelenterata or Radiata to represent a taxon uniting the two. While most workers in the field and most recent phylogenetic analyses reject this grouping, there have been some attempts to resurrect it based on data from the fossil record and on the inherent instability of many genome-scale phylogenies (see Box 11.2).

Cnidarian body organization

Cnidaria is the phylum that includes jellyfish, corals, sea anemones, hydroids, and many others. It is a diverse group of animals of significant ecological importance in the marine realm. Their importance lies both in the position they hold in the food chain as predators and as prey, and in the fact that one group of cnidarians, the reef-building corals, create an environment that is home for thousands of other species. Almost all cnidarians live in the sea, but there are several groups that have representatives in fresh water as well. Cnidarians can often be very colorful. They display a range of different body shapes and sizes and are often graceful in their movements as they drift in the currents or wave their arms (tentacles) around looking for food items.

The two post-embryonic germ layers in cnidarians are the internal **gastrodermis**, derived from the endoderm and external **epidermis** derived from the ectoderm (Figure 9.1). There is an intermediate acellular layer called the **mesoglea** between the two epithelial layers. The mesoglea varies in thickness, sometimes being a mere thin separation between epidermis and gastrodermis, and sometimes forms much of the bulk of the animal. The mesoglea is composed of largely acellular connective tissue embedded in a gelatinous matrix. The mouth leads to a sac-like digestive cavity known as the **gastrovascular cavity** or **coelenteron**. The digestive system is simple, usually undifferentiated, and undivided, but there are cases, especially in larger species, where it is functionally divided into a few regions.

Cnidarians all display radial symmetry at least to some extent, but as we said in Chapter 8, radial symmetry is rarely perfect. In cnidarians it is expressed as repeats of units around the central oral–aboral axis, with the repeats occurring usually four, six, or eight times in different taxa. Among the repeating

Organismic Animal Biology. Ariel D. Chipman, Oxford University Press. © Ariel D. Chipman (2024). DOI: 10.1093/oso/9780192893581.003.0009

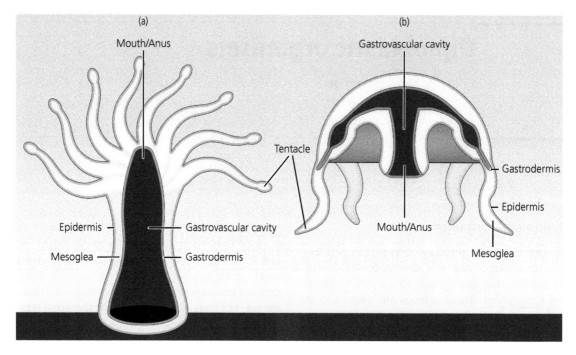

Figure 9.1 Body organization in Cnidaria: (a) The structure of a cnidarian polyp. (b) The structure of a cnidarian medusa.

structures found in some cnidarians are the mesenteries, thin septa dividing the coelenteron into separate cavities. The number and structure of the mesenteries is one of the characters used to differentiate taxa. In other cases, the repeating units may be sensory structures, tentacles, or gonads.

The unique character of cnidarians is the presence of stinging cells, **cnidocytes** (also known as **nematocytes**), which give the phylum its name (Figure 9.2). These are mostly found on tentacles that surround the mouth opening. The cnidocytes are highly specialized cells that deliver microscopic darts armed with venom into the body of potential prey. Inside each stinging cell is a capsule called the **nematocyst**, into which are packed a barb and a long thread. When the cnidocyte is activated, the capsule is opened, releasing the barb and thread at high pressure, sufficient to puncture the integument of the prey and inject the paralyzing venom into its body. The release of the nematocyst is driven by osmotic pressure and is one of the fastest-known biological processes. While most cnidocytes are used as hunting structures and act by penetrating as described, there are other types of stinging cells in cnidarians. **Spirocytes**, found in some cnidarian taxa, cling to

the prey and prevent it escaping. Burrowing cnidarians have an additional type of cnidocyte that help produce the tube in which they live.

The venom is composed of a cocktail of different proteins, optimized to target the nervous system or muscles of each species' specific prey items. In addition to injecting venom directly into prey via the nematocysts, cnidarians can also deliver venom by releasing it directly into the surrounding water, so that it is taken up by the prey through existing openings or wounds. The main purpose of the stinging cells is hunting, and nearly all cnidarians are carnivores. However, stinging is also useful as a defense mechanism against large potential predators, as anyone who has been stung by a jellyfish knows, and is used by some cnidarians to combat other individuals for space on the sea floor.

Cnidarians can exist in two main forms: the **polyp** and the **medusa** (plural **medusae**). Polyps are usually sessile or occasionally sedentary. The mouth points upward and is surrounded by a ring of cnidocyte-bearing tentacles that hunt food and bring it to the mouth. The mesoglea in most cases is relatively thin, giving polyps a general goblet-like shape. Medusae in contrast, are usually planktonic,

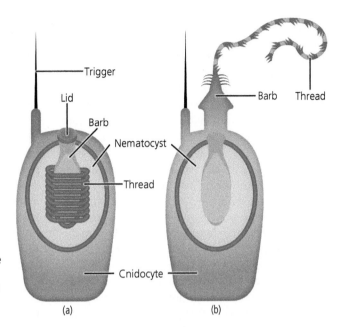

Figure 9.2 The structure of a cnidarian cnidocyte: (a) The structure of a cnidocyte before the release of the nematocyst. (b) A cnidocyte after the nematocyst has been released, showing the barb and venom-carrying thread, which penetrate the prey's integument.

and swim in the water column with their mouth oriented downward. The mesoglea is robust, giving a more bowl-like shape. The tentacles normally surround the margins of the bowl, and there are often additional arms surrounding the mouth. Some groups of cnidarians have both forms and some only one, as detailed below. When they are present, the medusae are the reproductive stage.

Cnidarian life history

There are many diverse life history modes in Cnidaria. These can be differentiated based on the presence or absence of the medusa stage, on whether the animals are **solitary** or **colonial**, and on the level of differentiation between polyps in colonial forms (see Chapter 10 for a discussion of coloniality). We will now detail three examples of life histories that cover some of the diversity found in cnidarians. These specific species were chosen because they have been well studied and are often used as examples in textbooks and in teaching labs, so the reader is likely to encounter them in other contexts. There are many variations beyond these examples. We will cover a few more examples later in this chapter, but with less detail.

The starlet sea anemone *Nematostella vectensis* has recently become the main lab organism for studying cnidarian development and other aspects of cnidarian biology. Its genome was fully sequenced in the early 2000s making it one of the first species sequenced when comparative genomics began. It lives in brackish water in estuaries and lagoons along the Atlantic coasts of North America and Western Europe. The life history of *Nematostella* is simple and includes a polyp stage and no medusa stage. It is solitary in all stages. The fertilized egg develops into a **planula** larva that swims in the plankton using cilia. The planula swims for a few days until it settles and metamorphoses into a polyp. Once the polyp reaches a size of a few centimeters, it achieves sexual maturity. It can produce either male gametes (sperm) or female gametes (ova) that are released into the water, where they meet and fuse to start the life cycle anew. *Nematostella* can also reproduce asexually through budding. In this case, the polyp pinches off a small fragment at its aboral end, by transverse fission, and this tissue piece then forms a head and a mouth at its oral end and continues its life as a solitary polyp.

The moon jellyfish *Aurelia aurita* is one of the most common jellyfish in the northern Atlantic. Its life history includes both a polyp and a medusa, and it is solitary in both stages. Like *Nematostella*, sexual reproduction creates a planktonic planula. The planula settles to form a feeding polyp. The polyps are small, rarely reaching more than 1 centimeter in size. Once it reaches this size, the polyp begins a process of **strobilation** wherein the polyp divides transversely into a series of plate-like structures called **ephyrae** that detach from the polyp and become juvenile medusae. When the medusae mature, they produce male or female gametes that are released into the water to meet and start a new life cycle.

Obelia is a genus of cnidarians with an almost worldwide distribution. *Obelia* has both a polyp and a medusa, but unlike the examples we have seen so far, the polyp stage is colonial. The early stages of the life cycle from fertilized egg to polyp are the same as in the previous species. However, the polyp stage undergoes asexual reproduction, by a lateral budding process. The nascent buds do not detach, and the animal develops into an ever-growing, bush-like colony of attached polyps, all descendants of a single founder. When the colony reaches a certain size, and the environmental conditions are suitable, some of the polyps differentiate into reproductive polyps—**gonozooids**. Medusae develop inside the gonozooids and are then released to carry out the sexual reproductive phase. *Obelia* colonies show division of labor, with different polyps carrying out different roles for the benefit of the colony. We will discuss division of labor in colonial organisms in more detail in Chapter 10.

We can draw some generalizations from the examples above: Medusae are always sexually reproductive. In species that produce medusae, there is an alternation between a medusa stage and a polyp stage (with a few exceptions). Medusae are generally solitary, but polyps can form colonies through asexual reproduction. Different polyps in a colony can have different roles, usually linked with differences in polyp morphology.

Cnidarian feeding and digestive system

Nearly all cnidarians are predominantly carnivorous. Polyps wave their tentacles and catch small animals, then bring them into their digestive system. Some medusae are capable of catching large animals, sometimes even larger than themselves, using their long oral arms. As a rule, cnidarians are opportunistic predators and will feed on whatever comes in contact with their arms or tentacles. However, their venom components are optimized for specific prey. The cnidarians that are most harmful to humans are those that typically prey on fish, and therefore have venom components that preferentially target vertebrate nervous systems.

Prey that is brought into the coelenteron is digested by enzymes that are secreted by the gastrodermis. In large medusae, the digestive system extends to radial canals that spread into the mesoglea, thus providing nourishment to all of the animal's tissues. In some large polyps there is some regional differentiation in the digestive system, but not much is known about the roles of the different regions.

Muscles and movement

Cnidarians move their arms and tentacles using epithelial muscles of endodermal or ectodermal origin. Medusae locomote by synchronized pulsing movements of radially arranged muscles, also of epithelial origin. These muscles work against the gelatinous tissue of the mesoglea. This generates a downward directed jet of water that allows the medusa to move up or down through the water column. With few exceptions, medusae are not capable of directional movement. On the other hand, polyps tend to be attached to the substrate or embedded in a mineral matrix and are incapable of any locomotion. There are, however, a few exceptions to this rule. Some polyps are capable of bending their body rapidly, causing them to "hop" short distances. This is normally an emergency response to an imminent predation threat. Other calcified polyps (solitary corals) are capable of creeping slowly across the surface. They do this by inflating themselves with water and pushing against the substrate.

Nervous system

The nervous system of cnidarians consists of a diffuse nerve net. There are no **ganglia** and no central processing region or brain. Their sensory organs are mostly simple, consisting of gravitational sensors (**statocysts**) and chemical senses. In most medusae, the sensory organs are concentrated in structures called **rhopalia**, situated around the margins of the contractile bell. The tentacles have touch receptors that facilitate grabbing prey once it has made contact with the tentacles. A notable exception to the simple sense organs of cnidarians are the box jellyfish (see below), which have evolved camera-type eyes that allow them to identify potential prey and actively capture it.

Cnidarian diversity and taxonomy

The phylum Cnidaria includes an estimated 10,000 species. Under Linnaean taxonomy, it is divided into five classes. The first major division of Cnidaria is into Anthozoa and Medusozoa (see Figure 9.3). Anthozoa includes a single class, which does not have medusae at any stage. Medusozoa includes all of the remaining classes that have a medusa stage in their life history. There are, however, some members of Medusozoa who have secondarily lost the medusa stage. The lack of a medusa stage is generally considered to be the ancestral state for Cnidaria, with the medusa probably evolving once, in the common ancestor of Medusozoa (although there is some debate about this question).

Anthozoa (literally, flower animals) includes two subclasses Octocorallia and Hexacorallia that are differentiated by their mode of radial symmetry: octoradial with eight mesenteries in the former and hexaradial usually with a multiple of six mesenteries in the latter.

Octocorallia includes the soft corals, sea pens, gorgonian corals, and their relatives. With very few exceptions, they are colonial, but with a relatively low level of differentiation into different polyp types. The highest level of differentiation is found in the sea pens, which have separate stalk polyps and branch polyps.

Hexacorallia includes Actinaria, the sea anemones (to which *Nematostella* belongs) and Scleractinia, the stony corals. Sea anemones are found in diverse sizes, from small anemones of under 1 centimeter, to robust species reaching over 10 centimeters in diameter. They are almost always solitary, with the few colonial species exhibiting no polyp differentiation.

The stony corals are arguably the most ecologically significant members of Cnidaria, and possibly the most ecologically significant marine organisms. The individual polyps are normally very small, often under 1 millimeter, but they live in large colonies with hundreds or thousands of individuals, living in a shared secreted calcium carbonate external structure. The mineralized structures of the corals form reefs, which are biodiversity hotspots. The corals themselves—together with other sessile organism growing on or among them—serve as a home for numerous additional species belonging to all marine taxa. The polyps of a coral colony are connected by a network of tubes within the mineralized skeleton, the **coenosarc**, which allows the transport of nutrients between members of the colony. An exception to the normal structure of corals are the plate or mushroom corals, which are solitary large polyps, each embedded in its mineralized skeleton. As mentioned above, these corals are also unusual in being capable of locomotion. Most corals have endosymbiotic photosynthetic algae known as zooxanthellae or dinoflagellates. The algae presumably supplement the nutrient

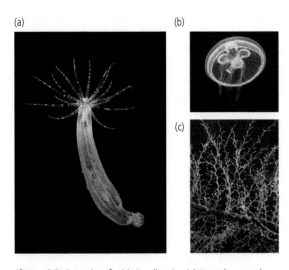

(a) (b) (c)

Figure 9.3 Examples of cnidarian diversity: (a) An anthozoan, the starlet sea anemone *Nematostella vectensis*. (b) A scyphozoan, the moon jellyfish *Aurelia aurita*. (c) A colonial hydrozoan, *Obelia geniculata*.

Source: a: Photo supplied from Shutterstock: Paul R. Sterry (https://www.alamy.com/stock-photo-starlet-anemone-nematostella-vectensis-79933516.html?imageid=06EB9105-35FC-4799-89D2-364EFE5EBA15&p=14455&pn=1&searchId=54e8bc177b736cfd64f69e38b94de620&searchtype=0); b: Photo supplied from Shutterstock: Vladimir Wrangel (https://www.shutterstock.com/image-photo/moon-jellyfish-aurelia-aurita-aquarium-236819803); c: Photo supplied from Shutterstock: Paul R. Sterry (https://www.alamy.com/stock-photo-obelia-geniculata-79935070.html?imageid=E47D7F4FC1A2-491F-84FA-8F836B3FC815&p=14455&pn=1&searchId=7b0786fe8e279642803f02ad47d1e1e8&searchtype=0).

input of the corals through photosynthesis. Corals can live after they have lost their endosymbionts, a phenomenon known as coral bleaching, which is caused by stress or environmental degradation. Nonetheless, bleached corals are unhealthy, and their lifespan is severely diminished.

Within Medusozoa, the best-known class is Scyphozoa (literally cup animals) the classic jellyfish. The moon jellyfish *Aurelia* is a member of this class and displays its typical life history. The medusa stage tends to be dominant both in size and in diversity. Indeed, for many species, almost nothing is known about the polyp stage, due to the miniscule size and difficulty in observing the full life cycle. Scyphozoan medusae can reach very large sizes, with bell diameters reaching up to 1 meter and tentacles reaching up to 10 meters. The medusae typically have a mouth cone, a structure unique to this class. They exhibit tetra-radial symmetry with the repeating units including gonads and rhopalia.

Hydrozoa is structurally the most diverse class of cnidarians, with a wide range of life histories and colony structures. The simplest forms are exemplified by the common freshwater species *Hydra*, which is a solitary polyp with no medusa stage and no larval stage. However, most hydrozoans do have a medusa stage, which is known as a **hydromedusa**. These tend to be much smaller than scyphozoan medusae (usually just 1–5 cm) and can be distinguished from them by a number of morphological differences. Hydrozoa includes many examples of colonial species often displaying a diversity of polyp types within a colony. *Obelia* is an example of such a species. The most extreme examples of differentiated colonies are found within Siphonophora. This is a group of pelagic colonial hydrozoans, that can include reproductive polyps, feeding polyps, protective polyps, structural polyps, and polyps that inflate and fill with air to serve as floats for the entire colony. Examples of siphonophores include the Portuguese man-o-war (*Physalia*) and by-the-wind sailors (*Velella*) as well as some of the longest known colonial organisms. One last hydrozoan group worth mentioning is the milleporine corals or fire corals. These colonial hydrozoans secrete a mineral skeleton, like the stony corals, but have a very different colony shape, usually looking like fans or fronds.

The third medusozoan class is Cubozoa, the box jellyfish or sea wasps. This is a small group of active predators, often with extremely potent venom, which may be lethal to humans. In contrast to all other cnidarians, they have well-developed sense organs, including four stalked eyes with a lens, and are capable of directional movement when hunting or navigating through mangroves. Cubozoans are also unusual in that some species have lost the polyp stage, and the planulae develop directly into medusa.

The somewhat obscure Staurozoa, or stalked jellyfish, forms the last Linnaean class within Cnidaria. There are about 50 species in this group, comprising small polyps with a morphology reminiscent of medusae that lived attached to a substrate.

Finally, the most derived and unusual cnidarians are the parasitic Myxozoa. These are highly simplified and reduced animals that mostly parasitize fish and have been known to cause significant damage to commercial fisheries. Myxozoans were only linked to Cnidaria recently following molecular work and are probably the sister group to all medusozoans. They possess structures known as **polar capsules** that are morphologically related to cnidocytes. Myxozoans have a complex life cycle that does not include a polyp or a medusa stage but does include a stage where they superficially resemble single-celled organisms. We will discuss simplification following parasitic lifestyles in Chapter 14.

Evolutionary history

Cnidarians are an ancient animal group. According to molecular clock estimates, the two main subphyla split about 700 million years ago, long before they appear in the fossil record. The cnidarian fossil record extends back to the Ediacaran, about 570 million years ago, with several fossils interpreted as putative scyphozoan medusae. The mineralized members of Cnidaria have an excellent fossil record, with several extinct groups of corals being common fossils throughout the Paleozoic. The modern Scleractinia appear in the fossil record and diversify in the Jurassic.

The emergence of *Nematostella vectensis* as a lab research animal has made it possible to address many questions about the transition from radial to bilateral symmetry and the origin of mesoderm. The general consensus has been that radial symmetry and lack of mesoderm are primitive characters for this phylum. However, molecular developmental studies show that this picture may be an over-simplification. Cnidarians have the genetic components needed to generate mesoderm, and there may be a regionalization of proto-mesoderm in cnidarians. Furthermore, it has been shown that anthozoans have an additional previously unrecognized body axis, the directive axis, raising the intriguing suggestion that they might be secondarily radial and might share a bilateral common ancestor with bilaterally symmetrical animals (see Box 11.2).

Ctenophore body organization

The second diploblastic phylum is Ctenophora, the comb jellies or sea gooseberries. Ctenophores are not nearly as diverse or common as cnidarians and are not as well known to the layman. While the two phyla share a superficial similarity, there are many important differences. Ctenophores display radial or biradial symmetry. Biradial symmetry is radial symmetry along the oral–aboral axis that is bisected by a bilateral plane. They have no nematocytes and no stinging tentacles. Instead, their tentacles possess "sticky" cells called **colloblasts**. The unique feature of the group, which also gives its name, is the presence of comb rows (**ctenes**) along the external surface of the body, running parallel to the oral–aboral axis (Figure 9.4). These are elongate structures bearing numerous cilia that beat in unison. All extant ctenophores have eight comb rows. The middle layer of the ctenophore body is a cellular mesenchyme or mesoglea that is derived from the endoderm. Beyond these characteristics, ctenophores are very often transparent or at least translucent. Many species have specialized cells that emit light when disturbed. Their transparency and constantly beating ctenes making them mesmerizing to watch. It is worth noting that being transparent is not a trivial feat. It is not enough to be unpigmented. There has to have been selection for a very specific composition and arrangement of tissues to allow light to pass through the animal, indicating that transparency has an important role in their survival.

Ctenophore life history

Ctenophores have simple life histories with direct development from a fertilized egg. No colonial forms are known. They reproduce sexually, with most species being hermaphrodites, either simultaneously or sequentially. Some have the ability to self-fertilize. The gonads, male or female, lie under the comb rows. Some ctenophores are known to have high regenerative capacities, and there are a few known cases of reproduction through budding or splitting (mostly in the bottom-dwelling Platyctenida).

Ctenophore feeding and digestive system

All ctenophores are predatory active hunters. Most catch prey using tentacles covered with colloblasts that eject a sticky substance to ensnare prey. Others have specialized structures to funnel prey into their highly extendable mouths. There is even one group of ctenophores that are specialized in eating other ctenophores. While ctenophores do not possess

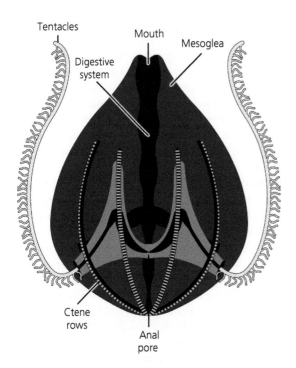

Figure 9.4 Body organization of a generalized ctenophore.

stinging cells, there are a few known species that feed on cnidarians and manage to sequester their prey's cnidocytes and use them themselves. This is known **kleptocnidism**. Ctenophores are generally considered to have a simple gut with one opening, like cnidarians. In fact, careful observation shows that the digestive system has an aboral opening as well, in the form of two small **anal pores**.

Muscles and nervous system

Locomotion in ctenophores is mostly through the synchronized beating of the ctene rows. Ctenophores are considered to be the largest animals that swim through ciliary action. Most ctenophores are pelagic, with one group, Platyctenida, being benthic crawlers, often associated with echinoderms. In addition to the comb rows, ctenophores also use muscles. These are endoderm-derived muscles, which may have evolved independently of the muscles of all other animals (see Box 9.1).

Ctenophores possess nerve cells of several types, arranged in a nervous system that is more complex than the cnidarian nerve net, but still cannot be counted as a central nervous system. They have minimal sensory organs, mostly statocysts, located in the aboral pole, and chemical and pressure receptors, concentrated around the oral opening. There are no visual organs known, but ctenophores are known to respond to the daily light cycle.

Ctenophore diversity

Ctenophora probably includes several hundred species, although only about 100 have been described. Many species live in deep-water species and are difficult to collect. Beyond that, their fragile structure makes them difficult to preserve for taxonomic studies. The taxonomy of Ctenophora is not very well resolved and there is no consensus on which characters should be used for defining the different taxa within the phylum. Some ctenophores have tentacles and others don't, and this was used as a key character dividing ctenophores into two classes. The presence of tentacles is now believed to be a labile character, so this division is not widely accepted. Some of the main accepted groups of ctenophores are listed below:

Beroidae—pelagic ctenophores with no tentacles and an expanded mouth that can engulf large prey. They have lost many other characters, as well as the tentacles, including regenerative capabilities. Many beroids are specific predators of other ctenophores.

Platyctenida—The only benthic group of ctenophores. Platyctenids have flattened bodies and often live on other benthic organisms, such as corals or echinoderms. They possess comb plates during embryonic development, but lose them as adults. Platyctenids are the only ctenophores that normally reproduce through asexual development.

Cydippida—Pelagic ctenophores with long tentacles. Often with a flattened body. Considered to be the most conservative group. The cydippid body plan is often given as the ancestral ctenophore body plan.

Lobata—The most familiar ctenophores, including the well-known *Mnemiopsis*. Relatively short tentacles and typical oral lobes that are used for prey capture.

Ctenophore evolutionary history

Based on their phylogenetic position (see Box 9.1), ctenophores are expected to have evolved very early in the history of animals. Some enigmatic Ediacaran fossils have recently been interpreted as ctenophores, providing evidence for their early appearance. Throughout the Cambrian, there is much higher diversity of animals believed to be ctenophores, including several examples of armored ctenophores. Apparently, the ctenophores found today represent a very small subset of ancient ctenophore diversity. This idea is supported by molecular clock analyses, which place the most recent common ancestor of modern ctenophores much later than the first appearance of the group as a whole. Generally speaking, the relationships of ctenophores to other animals have always been difficult to resolve using morphological data, due to their unusual set of characters. The advent of genetic tools has not helped resolve the problems, as they seem to have several unusual aspects in their molecular evolution as well.

Box 9.1 The debate over the earliest branching animals

Until relatively recently, most zoology textbooks took for granted the fact that sponges were the most "primitive" animals and that they were the sister group to all other animals. This long-accepted fact was challenged in the early 2000s with the increased use of genomic data to reconstruct the relationships among animal phyla, and the sequencing of the first ctenophore genomes. A number of papers consistently recovered a phylogeny in which Ctenophora is the earliest branching metazoan lineage. This result was surprising not only because ctenophores had traditionally been allied with cnidarians under Coelenterata, but because it prompted a significant rethinking of the early evolution of a number of characters that define animals.

The idea of sponges being primitive was supported by the fact that they lack muscles and lack a nervous system, in contrast with all other animals that have both. If ctenophores, with their muscles and nervous systems, are indeed an earlier branching lineage that would require one of two scenarios to be true: either sponges had secondarily lost muscles and nervous systems or ctenophores had evolved muscles and nervous systems convergently.

The debate that arose following this suggestion has been raging unabated for nearly two decades. In fact, it has developed into separate debates. On the one hand is the phylogenetic debate, with new datasets and new analytical tools being used to provide a better supported tree indicating one or the other phylum as the sister to all other animals. On the other hand is the evolutionary debate trying to provide evidence supporting the homology or lack of homology between muscles and the nervous system in ctenophores and other animals. At the time of writing (early 2023), it looks like there is more support for Ctenophora being the sister group to all other extant animals, based on a larger dataset supporting that phylogeny and on evidence that suggests the ctenophore nervous system is indeed convergent to that of all other animals. However, there is no consensus yet and the debate is probably far from over.

CHAPTER 10

Colonial organisms and complex life cycles

What is a colony?

In our discussion of cnidarians in the previous chapter, we saw that many species live in colonies. Colonial organisms are found not just among cnidarians, but among many other taxa, as we shall see when we continue our survey of animal phyla. We must therefore wonder why this mode of living is so common. Let's start by defining a colony; a colony is a group of organisms of the same species, living together in a single location, often to the exclusion of other species. This is a very broad definition that encompasses many types of colonies. Members of a colony may interact minimally, or even not at all, or they can work together for the benefit of the colony. They can be closely related or very distantly related. The members of a colony can be permanent, or the colony can have a constant influx and outflux of individuals. Groups of large mobile animals are usually known as herds, flocks, swarms, or any one of many other terms. The same principles that apply to colonies are relevant for them as well.

Given this diversity in colony types, we can ask what colonies have in common that gives them an evolutionary advantage and has caused them to evolve so many times. One important advantage is what has been called the dilution effect. When there is a limited number of predators, any individual gains an advantage by being in close proximity to other individuals who can serve as potential prey, thus reducing the individual's chances of being the one targeted by the predator. Being in a large group also increases the chance that some member of the group will detect the predator in advance and "sound the alarm." In addition, being together in

a group of conspecifics reduces the energetic cost of looking for a mate and increases the chances of finding one.

Colonies can form when an important resource is rare. This can be a food resource or an environmental resource, such as suitable living substrate. There is often an advantage to individuals sharing the resource, over expending energy on competition over the resource, even if it means living at higher densities. This is especially relevant in sessile or sedentary organisms, where movement between suitable sites is impossible or limited. Furthermore, in some cases, members of a colony will collaborate in obtaining or utilizing a resource, giving a further advantage to living in a colony.

Another advantage, which stems from the ones already mentioned, is the advantage of size. A larger colony is more resistant to predation, is able to obtain food and other resources more easily, and is energetically more efficient. Once a colony forms initially, it will tend to get larger, as long as sufficient resources are still available.

There are two main ways colonies can come together. One is through aggregation, where members of a species transmit information about a potential colony site, and other individuals use this information to join the colony. In this type of colony, the genetic relatedness among members of the colony is unimportant. Any individual who has the information about the colony can join it. At the other extreme are **clonal colonies**. These colonies are started by a founder individual, which then undergoes asexual reproduction or cloning to generate additional, genetically identical individuals who form the colony. This type of colony is what

Organismic Animal Biology. Ariel D. Chipman, Oxford University Press. © Ariel D. Chipman (2024). DOI: 10.1093/oso/9780192893581.003.0010

we find in cnidarians and in several other sessile marine species. The individuals of such a colony are usually known as **zooids**. An intermediate stage between these two is a **breeding colony**. In this type of colony, individuals reproduce in a certain area and their offspring tend to return to the same spot to reproduce when they are sexually mature. In this case, individuals of the colony are genetically related due to the colony's history, but they are not identical.

Many colonial organisms exhibit some level of division of labor, where individuals specialize in certain tasks that are necessary for the survival of the colony as a whole. The division of labor can be transient, meaning that morphologically similar individuals carry out different tasks, but these tasks can change over time. The division of labor can also be fixed, with specialized individuals being morphologically different and optimized for one role. This type of division of labor with morphological differentiation is found mostly in cnidarians and is somewhat rarer in members of other phyla. Division of labor is taken to an extreme in some hydrozoan cnidarians, where some individuals specialize for reproduction and other member of the colony do not reproduce at all (Figure 10.1). There is an advantage to specific reproductive individuals in clonal organisms, where all members of the colony share the same genome.

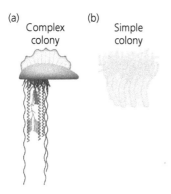

Figure 10.1 Types of colonies: Two colonial cnidarians with different degrees of organization. (a) A colony of the hydrozoan *Physalia,* the Portuguese man-o-war. The colony is composed of at least five types of polyps, with only one type being involved in reproduction. (b) A simple anthozoan colony composed of only one type of polyp. All polyps are sexual and can reproduce.

If you have been reading this book in sequence, much of the preceding discussion may sound familiar. We presented many similar considerations when discussing the origin of multicellularity in Chapter 5. Indeed, there are many parallels between the evolution of multicellularity and the evolution of coloniality, and especially complex coloniality. This can be seen as part of a general trend toward increasing complexity in evolution, which we will discuss at length in Chapter 19.

The seminal book *The Major Transitions in Evolution* by John Maynard Smith and Eörs Szathmáry, published in 1995, outlines a sequence of increases in complexity over evolutionary time. Maynard Smith and Szathmáry include the evolution of multicellularity as one of these transitions. The next transition they list is the evolution of **eusociality**. This is a special case of coloniality that involves not only the existence of reproductive and non-reproductive individuals, but also cooperation among members of the colony for the benefit of the reproductive individuals and their offspring. Hydrozoan colonies do not fall under this definition because of the lack of cooperation. Eusociality is mostly found in insects, with the most famous examples being termite, ant and bee colonies. "Major Transitions" notwithstanding, we can identify a series of intermediate stages leading to eusociality, with examples including bumblebees and wasps that exhibit transient or partial eusociality. Partial eusociality is also found in some spiders, while the naked mole rat is the only mammal to show full eusociality.

If there are such parallels between multicellular organisms and colonial organisms, where do we draw the line? What prevents us from referring to a coral colony or a Portuguese-man-o-war as a single organism? Can we call a beehive a single "super-organism"? The discussion is largely semantic and depends on how we define an organism. Under one definition we could include colonial organisms under the same heading as solitary organisms. Under a different definition, we would draw a sharp line between them. We believe that the place to draw that line is at an individual's **ontogeny**. Even within a colonial organism, each individual making up the colony starts out as a fertilized egg or as a bud from its parent. The sequence of changes from that bud or egg and to the adult is its ontogeny, and it is what makes the individual a separate organism within a colony.

Life cycles

Many animals have a **simple life cycle** with **direct development** (Figure 10.2A). Following embryonic development, the animal hatches from an egg or is born from its mother. It then grows following the species' ontogenetic path. It reaches sexual maturity, reproduces, and dies. Conversely, many other animals have more complex, **biphasic life cycles** (Figure 10.2B). These animals have very different juvenile stages and adult stages, punctuated by a short period of transition called **metamorphosis**. The juvenile of animas with biphasic lifestyles is usually referred to as a **larva** (plural larvae). The adult and the larva often have different ecologies, utilize different resources, and have different roles in the success of the organism.

A step up in the complexity of life cycles involves **alternation of generations** (Figure 10.2C). The organism alternates between a sexual generation and an asexual generation. The asexual generation often reproduces clonally to increase its numbers, while the sexual generation is often responsible for dispersal. Note that in alternating generation, the individual organisms are separate between generations, as opposed to biphasic life cycles, where it is the same individual before and after metamorphosis. In medusozoan cnidarians, the medusa stage and the polyp stage are separate animals, whereas in anthozoans, the planula larva metamorphoses to give a polyp, that is the same individual animal.

Many cnidarians combine complex life cycles (with or without alternating generations) with complex colonies with division of labor. This variability in life cycles underlies much of the morphological and ecological diversity of Cnidaria. Despite the conserved and simple body plan of the zooids, cnidarians can be found in a range of higher-level organizations. We can say the same about other marine phyla we will meet further on.

Types of larvae

A biphasic life cycle with separate larvae and adults is very common in marine animals, and especially in sessile or sedentary animals, where the larva is the only mobile phase. We can broadly divide larvae into two main types based on where they obtain their nourishment during the larval phase: **planktotrophic** larvae, which trap and feed on small

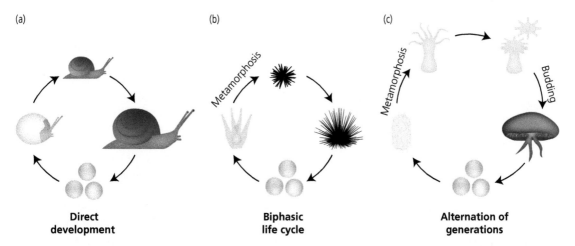

(a) **Direct development** (b) **Biphasic life cycle** (c) **Alternation of generations**

Figure 10.2 Different types of life cycles. (a) In direct development, exemplified by a snail, the juvenile that hatches from the egg is similar to the adult. (b) In a biphasic life cycle, exemplified by a sea urchin, the larva that hatches from the egg is very different from the adult, and the animal's life cycle includes a dramatic metamorphosis in which the larva transforms into a juvenile that is similar to the adult. (c) In a life cycle that includes an alternation of generations, exemplified by a jellyfish, there are two separate phases. Only one phase involves sexual reproduction, whereas the other phase may include asexual reproduction, such as budding.

planktonic organisms, and **lecithotrophic** larvae, which survive on stores of internal yolk provided by the mother in the egg (or ovum). Both types can be found in closely related species, and the transition between the two occurs easily over evolution.

Beyond this broad ecological division, larvae can come in many different types. As a rule, they tend to be morphologically simple, and they do not display the level of diversity and specialization found in the adults of the species. Cnidarians have simple, flat, ciliated **planula** larvae. The **trochophore**-type larva (Figure 10.3) is found in a number of marine phyla. It is characterized by a medial **ciliary band** that provides locomotion and by an **apical tuft** of hair-like structures that form the basis for the adult central nervous system. Many marine snails have an advanced larval stage that follows the trochophore larva, known as the **veliger** larva. Sea stars, sea urchins, and their kin have slightly more elaborate larvae, often with a simple skeleton, and rudimentary digestive systems. There are many examples in evolution where the larva is lost in some lineages. Sometimes the organism's ontogeny completely skips the larval stage, while in other cases, the larval stage is incorporated into embryonic development and occurs within the egg.

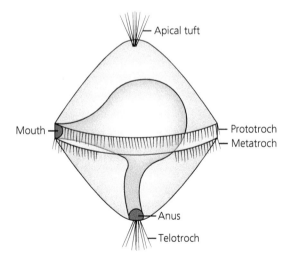

Figure 10.3 The trochophore larva, a type of larva found in several phyla.

Metamorphosis

Metamorphosis is a dramatic transition in body plan over a relatively short time. The emphasis here is on *relatively* short, since despite common perceptions, metamorphosis is not an immediate event, but a process. During metamorphosis, the organism undergoes both a morphological and an ecological transition. After metamorphosis, the organism's digestive system is likely to be completely different, and sense organs will be adapted to a different environment. Often, metamorphosis in marine animals is a transition between planktonic drifting to sessile living. Note that in most cases, metamorphosis does not signify the end of an organism's ontogeny, and growth can continue well after metamorphosis.

To understand the diversity of types of metamorphosis, let's look at a few familiar and less familiar examples, representing a wide taxonomic range. We'll start with perhaps the best-known example, that of a tadpole transforming into a frog. During metamorphosis, the tadpole's tail is resorbed, its limbs develop and emerge from the body wall, the skeletal system develops gradually as cartilage is replaced by bone, the jaws are transformed, and the gut is restructured so it can process animal material rather than the algae the tadpole fed on. Despite the many changes, this process is rather gradual and includes several distinct events that take place sequentially over a few days.

Another familiar example is that of a caterpillar metamorphosing into a butterfly (or a maggot metamorphosing into a fly). In insect metamorphosis, there is a quiescent period that take place inside the pupa. During this period the animal normally doesn't feed or move. While in the protective environment of the pupa, many tissues are broken down via a process of **histolysis**. In parallel, new tissues emerge from dedicated structures known as **imaginal discs**.

The trochophore larva is the first post-embryonic stage in many marine phyla, including several bristle worms (annelids—see Chapter 17). After swimming in the plankton for a few days, the larva begins to develop from its posterior side. Unlike the previous examples, metamorphosis does not involve the breakdown of any larval tissues, but an elaboration of the existing ones. The gut elongates and differentiates. The central nervous system is built upon the apical tuft, the ciliary band expands to give adult cilia, and the posterior develops into a growth zone. Some bristle worms live sedentary lives in the substrate or in a tube, and in these cases, metamorphosis leads to an ecological shift. However, others remain as swimmers or crawlers, and there is a smooth transition from the larval lifestyle to the adult.

Our final example is the sea urchin. Sea urchins belong to the echinoderms (see Chapter 26), a phylum of mostly sedentary marine animals. Their pyramid-like planktonic larva is called a **pluteus** larva. In many senses, sea urchin metamorphosis is similar to that of butterflies, despite radically different body plans and ecologies. As in the butterfly, metamorphosis involves the breakdown of almost all larval tissues. The adult body derives from small population of cells that start to develop inside the larval body. At metamorphosis, the adult rudiment breaks out of the larva, engulfs it, and resorbs it.

CHAPTER 11

An introduction to Bilateria

The route to bilaterality

Bilaterally symmetrical animals comprise the over-whelming majority of metazoan species, outnum-bering non-bilaterian animals 100 to 1. At higher taxonomic levels, most animal phyla are bilateral, with only four non-bilaterian phyla. All bilateral animals are grouped together under Bilateria, a monophyletic super taxon. The fact that Bilateria is monophyletic indicates that bilaterality evolved once early in the history of the animal kingdom. This is believed by most researchers to have been at or around the most recent common ancestor of Bilateria, but there is also evidence to suggest that bilateral symmetry may have evolved much earlier (See Box 11.1).

We can draw a hypothetical but likely scenario for the early evolution of Bilateria. The idea at the base of this scenario is that the driving force was the appearance of directional locomotion, as dis-cussed in Chapter 8. As movement became direc-tional, more sense organs were concentrated at the end of the body leading the movement—the ante-rior end. Together with an anterior concentration of sense organs, early bilaterians evolved an anterior processing center, the primitive brain. As the main reason for moving is to find food, the mouth grad-ually shifted anteriorly. Additionally, the digestive system became more complex, and the intake open-ing was separated from the waste removal opening, leading to the appearance of the anus. A diges-tive system with two openings is also known as a **through gut** and is one of the defining characters of Bilateria.

The concentration of sense organs, neural pro-cessing, and a mouth at the anterior of the body and the formation of a distinct head is known as **cephalization**. The extent of cephalization is vari-able among bilaterians, with some having clearly defined and separate heads and others having only some aspects of cephalization. Cephalization may have originated early in bilaterian history and been reduced later in some lineages, or the process may have initiated several times within Bilateria. The taxon believed to be the earliest branching within Bilateria, Xenacoelomorpha (see Box 11.2), has minimal cephalization, with no ante-rior brain, no anterior mouth, and no through gut.

Within Bilateria, there has been a recurring ten-dency for the anterior–posterior axis to become elongated. There are many possible reasons for such an elongation to occur: improved locomotion over the substrate or within a soft substrate, greater effi-ciency and differentiation of the digestive system, or a general outcome of the way many bilaterians grow through extension of the posterior. These long and thin animals are generally known as **worms**. How-ever, "worm" is an imprecise term with no phylo-genetic significance. Throughout this book we will come across many phyla that are known as worms (e.g., flatworms in Chapter 13, segmented worms in Chapter 17, and roundworms in Chapter 20) or that include animals classified as worms, but these phyla are not more closely related to each other than they are to non-worm phyla.

Mesoderm

So far, we have covered two germ layers: endo-derm and ectoderm (Chapter 7). The third germ

Organismic Animal Biology. Ariel D. Chipman, Oxford University Press. © Ariel D. Chipman (2024). DOI: 10.1093/oso/9780192893581.003.0011

Figure 11.1 Arrangement of the three germ layers in a schematic triploblast.

layer, the mesoderm, evolved early in the history of Bilateria. Animals with three germ layers—endoderm, ectoderm, and mesoderm—are known as **triploblasts** (Figure 11.1). All bilaterians are triploblasts and all triploblasts belong to Bilateria.

Mesoderm arises in embryonic development as part of the process of **gastrulation** (see Chapter 25 for a more detailed discussion). During this process, epithelial tissues sink into the embryo, and as they sink, they undergo a transition from two-dimensional tissues to three-dimensionally arranged mesodermal tissues. These are predominantly **connective tissues** of various types, which are made up of cells in a loosely ordered three-dimensional structure, supported by fibrous extracellular tissue. There are many types of connective tissue and not all conform to this description. They are normally found in the spaces between epithelial tissue layers. The appearance of connective tissue and of various support tissue derived from them, allowed animals to grow larger.

The appearance of mesoderm also allowed the elaboration of muscles. Recall that we met epithelial muscles in diploblastic organisms in Chapter 9. Mesodermal muscles have more complex organization and can form muscle bundles that provide more power. This is likely to be linked to the evolution of active directional movement, but it is difficult to say which came first, mesodermal muscles or directional movement.

In addition to support tissue and muscles, embryonic mesoderm gives rise to a wide range of structures. Connective tissues form most of the integument and of the digestive system, aside from the bounding epithelial layer, as already mentioned briefly in Chapter 7. Mesodermal tissues form many of the secretory glands associated with the digestive system and the integument. Mesoderm builds most of the circulatory system and excretory system in animals. Although embryonic mesoderm is not epithelial, it can give rise to adult epithelial structures, such as the epithelium that lines blood vessels. Finally, the gonads, which produce and store reproductive cells, are mesodermal in origin.

The diversity of Bilateria

There are three main branches to Bilateria. These arose from one major split, followed by a second split in one of the branches (see inner front cover). The first split is the Deuterostomia—Protostomia split. The names of these branches date back to the late nineteenth century and are based on shared aspects of early embryonic development, which we will discuss in Chapter 25. The deep split between deuterostomes and protostomes has been corroborated by almost all molecular analyses.

The larger of the two branches, both in terms of species number and in number of phyla, is Protostomia. In turn, it is divided into two. One branch is known as Ecdysozoa—the molting animals. As Ecdysozoa includes Arthropoda (insects, spiders, crabs etc.), which is the most diverse phylum, it is the largest in terms of species number. The second protostome branch is known as Lophotrochozoa or Spiralia. These terms are not quite identical, but for simplicity (and because it is easier to remember and pronounce) we will use Spiralia for the remainder of this book. Spiralia is more diverse in terms of number of phyla and body plans, including animals as diverse as snails and octopuses, flatworms, bristle worms, and other, but has fewer species overall.

The second branch of Bilateria is Deuterostomia. While it has much lower species numbers and includes fewer phyla, Deuterostomia is of special interest, since this is the branch that include the vertebrates, the group to which our own species belongs.

The first bilaterian

There has been an ongoing debate for over 20 years about what the first bilaterian might have looked like. This ancestral organism is referred to as **Urbilateria**. The crux of the argument is which characters found in Bilateria are homologous and therefore date back to the common ancestor, and which evolved convergently. One view reconstructs a very simple Urbilateria, with the bare minimum needed to be a bilaterian animal. Supporters of this view reconstruct a simple worm-like animal, with an anterior–posterior axis, a non-centralized nervous system or nerve-net, and a simple undifferentiated through gut, or a gut with a single opening. Some workers posit an alternative view of a complex Urbilateria, based on a series of similar structures across different bilaterians. They reconstruct it as having eyes and other anterior sense organs, a regionalized digestive system, a dorsal heart, a body made up of repeating segments, and a brain composed of three parts (a tripartite brain). (Figure 11.2).

Some aspects of developmental biology tend to support a complex Urbilateria, with shared components in some of the developmental pathways involved in generating the similar structures. However, as we discussed when we introduced the concept of homology in Chapter 4, superficial similarity often masks differences that tip the balance in favor of convergence. A broader analysis of the diversity of body plans in Bilateria, as well as data from the fossil record of early members of different phyla, suggests that at least some complex aspects of the body plan evolved independently in different lineages. This is probably the case for the segmented body plan (see Chapter 18). While the case is not clear-cut, the balance of evidence points toward a bilaterian ancestor that is closer to the depiction of the simple Urbilateria.

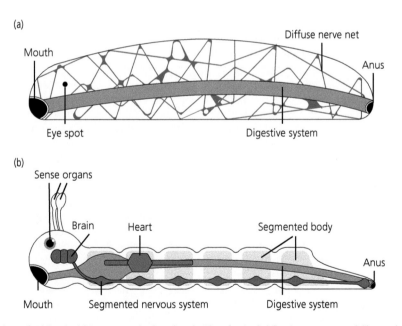

Figure 11.2 Two views of Urbilateria: (a) A reconstruction based on the idea of a simple bilaterian ancestor, much like members of Xenacoelomorpha. (b) A reconstruction based on the idea of a complex bilaterian ancestor, with many structures that are found throughout Bilateria.

Box 11.1 The origin of animal axes

The generally accepted paradigm about the evolution of animal symmetry sees radial symmetry as being primitive, since it is found in two of the non-bilaterian phyla, Ctenophora and Cnidaria. This view suggests that the anterior–posterior axis evolved in a radially symmetrical animal as part of the transition from a sessile mode of life to a mobile mode. The mouth is generally accepted to be homologous in all animals, and thus, the anterior–posterior axis can be expected to be homologous to the oral–aboral axis of non-bilaterians.

This view is being reconsidered in light of experimental evidence from anthozoan cnidarians. It turns out that anthozoans have two body axes, an oral–aboral axis like all cnidarians, and a second orthogonal axis that has been called the directive axis. The directive axis defines a bilateral symmetry in anthozoans. Based on developmental biology, the anthozoan directive axis shares some molecular traits with both the bilaterian dorso-ventral axis and the bilaterian anterior–posterior axis. The existence of bilateral symmetry within Cnidaria, which may be homologous with that of Bilateria, raises the intriguing possibility that the bilaterian–cnidarian ancestor may have been bilaterally symmetrical. This would push back the origin of bilateral symmetry to well before the origin of Bilateria. Under this scenario the radial symmetry of cnidarians is secondary to the original bilateral symmetry, with vestiges remaining in Anthozoa and all trace of bilaterality disappearing in Medusozoa.

This scenario is not so far-fetched if we imagine a placozoan-like cnidarian-bilaterian ancestor. Directional movement could have originally arisen in a diploblastic benthic organism with a simple gut with one opening and a nerve-net based nervous system. From this proto-bilaterian organism two lineages could have diverged. One lineage led to Cnidaria and maintained the diploblastic organization, while shifting to a mode of feeding that was not directional, concomitantly losing the obvious anterior–posterior axis but maintaining the genetic machinery in the directive axis. The second lineage led to Bilateria, evolved mesoderm and resulted in a xenacoelomorph (See Box 11.2) like bilaterian, in line with the simple Urbilateria scenario.

Box 11.2 Xenacoelomorpha as a proxy for the earliest bilaterian

Xenoturbella is an enigmatic worm-like animal found in deep cold ocean waters. Until very recently, it was known from only one species in one location, but additional species have since been found worldwide. Molecular phylogenies consistently ally *Xenoturbella* with another group of simple worms known as Acoelomorpha. Both *Xenoturbella* and acoelomorphs are characterized by a simple morphology, with a nerve-net based nervous system, no excretory system, no heart, and a digestive system with only one opening. The two taxa are therefore united under the phylum Xenacoelomorpha.

The phylogenetic position of Xenacoelomorpha has been under debate since it was officially defined. Most analyses place it as the sister group to all other bilaterians, branching off before the split between Protostomia and Deuterostomia. If this placement is current, members of the phylum provide a useful proxy for what the bilaterian ancestor probably looked like—very similar to the simple version of Urbilateria (Figure 11.1A). Under this interpretation, through-guts, excretory systems, circulatory systems, and other complex structures evolved late in bilaterian evolution, after Xenacoelomorpha branched off. Characters that xenacoelomorphs have in common with cnidarians, such as a nerve net and a digestive system with a single opening are homologous and represent the pre-bilaterian state.

Conversely, some of the first phylogenetic analyses of *Xenoturbella*, as well as more recent analyses of Xenacoelomorpha as a whole, place them as an early branching group within Deuterostomia. Under this interpretation, their simple body plan and lack of key organ systems are a secondary simplification; characters lost even though they existed in an early ancestor. If this is true, there is no a-priori reason not to assume homology between structures such as the heart or the brain between protostomes and deuterostomes, making the complex Urbilateria (Figure 11.1B) more likely.

The debate over the phylogenetic position of Xenacoelomorpha is another example of how the position of an obscure group of animals can have profound effect on our understanding of key transitions in animal evolution.

CHAPTER 12

Sensory systems

A wealth of information

The world is teeming with information that can be collected by the organism and used to improve its chance of survival. This can include information about the layout of the environment and its physical characteristics, information about other organisms in the environment and information about the organism itself. Over time, animals have evolved a wealth of tools to acquire and process this information. These tools are collectively known as sensory systems.

In this chapter, we will cover the main types of information available in the environment, also called different **sensory modalities**. We will discuss the common characteristics of the senses that read these modalities and give some examples. Many of the examples will be from mammals, which we can relate to our own subjective sensory experience. From these examples, we will try to move on to understanding sensory experiences that are very different from our own.

An introduction to nervous systems

Before discussing the sensory systems themselves, we will give an overview of the system that integrates and coordinates the information they collect—the nervous system. The basic unit of all nervous systems is the nerve cell or **neuron** (Figure 12.1). There is a diversity of sizes, shapes, and functions in different types of neurons, but they all share a key functional trait; the transferring of electrical pulses, known as **action potentials**, that constitute the information flow of the nervous system. They also share a number of structural similarities. Neurons are composed of a cell body or

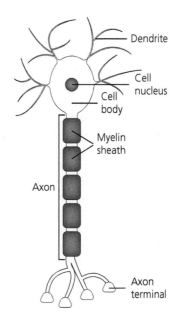

Figure 12.1 The structure of a generalized neuron.

soma, and two types of projections: **axons** and **dendrites**. The soma contains the nucleus and most other organelles and carries out many of the cell's biochemical processes, but it also acts as a miniprocessing center for electrical information signaling. A neuron may have one or many dendrites. The dendrites act as the electrical input unit of the neuron and transfer the signals they receive to the cell body. The axon is the electrical output unit, transferring electrical signals to other neurons and "exciting" them. The axon is usually covered by a **myelin sheath**—bands of a fatty material that improve conductance of the electrical signals.

Put very simply, the dendrites transfer action potentials in pulses to the soma, which weighs the

Organismic Animal Biology. Ariel D. Chipman, Oxford University Press. © Ariel D. Chipman (2024). DOI: 10.1093/oso/9780192893581.003.0012

different signals and integrates them. In response to the inputs, it can either "fire" sending a single electrical pulse along the axon, or not fire. In other words, every neuron receives a stream of inputs and responds with a single output. Neurons tend to act in networks, wherein every neuron receives inputs from many other neurons and transmits signals to yet many others. Every individual neuron receives and transfers signals in the form of individual pulses. However, the integration of pulses from many neurons in a network provides information that is quantitative and is about timing.

The type of neuron that interests us the most for this chapter is the sensory neuron, which receives inputs that are mostly external to the body. The rest of the chapter will give an overview of the structure and function of different sensory systems and will show that at the core of each of these systems are individual sensory neurons that receive information and respond with electrical pulses. The pulses from all sensory neurons are integrated by the nervous system. We usually differentiate between the **central nervous system** or CNS, which is where higher-level processing takes place, and the **peripheral nervous system** or PNS, which includes both sensory neurons and motor neurons, the latter exciting and activating the muscles.

In most bilaterians, there is a concentration of sensory organs in the anterior of the animal, the direction of motion and the direction where the most important information is to be found, as already discussed in Chapters 8 and 11. Many of these organs are paired, with one organ on the left and one on the right. This type of organization allows the organs to provide spatial information that can be interpreted by comparing the inputs from the two sides. Other sensory organs are more broadly distributed along the body, to provide information about the animal's entire surroundings.

Chemical reception

Chemical senses are the most ancient and probably the most diverse, both in the types of organs that mediate chemical reception (or **chemoreception**) and in the range of signals they can interpret. All organisms, from bacteria to plants, have at least some ability to detect specific chemical compounds and react to their existence. The basic unit of **chemosensory** organs is a population of receptors—usually proteins located on the cell membrane of a sensory neuron's input end—that can bind specific chemicals. Binding of the target molecule to the receptor leads to a cascade of interaction within the cell that ultimately leads to the cell producing electrical signals that stream along the axon and then excite other neurons (Figure 12.2).

The receptors on the sensory neuron can be very specific and bind only one sort of molecule, or they can be general and bind a range of similar molecules. An individual neuron may have only one type of receptor on its surface, or it can have a mix of receptors.

The two most familiar chemical sensory modalities in mammals are taste and smell. The sense of taste is mediated by five types of sensory neurons that each binds a broad family of molecules. Receptors on the sweet-sensitive neurons bind several different types of sugars (as well as artificial analogs of sugars such as saccharin). Sour receptors bind several organic acids. Salt receptors respond to cations of sodium and potassium in the environment. Receptors on the bitter sensing neurons bind a range of small alkaloids such as quinine. Finally, the recently recognized umami receptor binds amino acids such as glutamate. All of these receptors bind water soluble molecules that directly come in contact with the sensory cells that are present on the tongue.

The sense of smell (or olfactory system) is mediated by sensory cells, known as **olfactory cells**, which are present in large numbers in the nasal epithelia. Unlike the taste receptors, the receptors of the olfactory system are very diverse, with each type binding a relatively narrow range of small volatile compounds. Mammals can have up to 1000 different olfactory receptor-proteins, with humans having about 400. The sense of smell is combinatorial, so that different combinations of activated olfactory cells lead to the identification of different subjective odors. There are an estimated 1–2 million individual olfactory receptors in the human nose, allowing the identification of up to *one trillion* different odors.

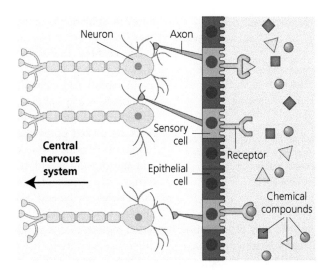

Figure 12.2 Schematic representation of the function of chemoreception: The right side of the image represents the outside world, with a diversity of chemical compounds suspended in the air or water. Specific receptors bind these compounds. The top and bottom receptors have bound their target chemical and the sensory cell to which they are attached transfers a neural impulse to an intermediate neuron and from there to the central nervous system.

Chemical senses provide animals with a lot of important information about their environment. They can indicate the presence of a food source and thus lead the organism to move in the direction where the concentration of a certain chemical is higher, or in the direction where the sensory input increases. This type of behavior where an animal moves in the direction of an increasing chemical gradient is known as **positive chemotaxis**. If the chemical indicates the presence of danger—a predator or a toxic substance—the animal will move in the direction of decreasing concentration of the signal. This is known as **negative chemotaxis**. Many species of animals use chemical senses as a means of communication with conspecifics. A substance emitted by one animal in order to transmit information to a different individual—usually of the same species—is known as a **pheromone**. The most common use of pheromones is in reproduction, where members of one sex use the pheromone to announce their readiness for mating and to attract members of the other sex. Because of the importance of reproduction for evolutionary success, detection of sex pheromones tends to be the most sensitive chemical sense in animals, with some male moths being able to detect pheromone molecules released by females several kilometers away.

Mechanical reception

The term mechanical reception or **mechanoreception** refers here to all the sensory modalities that give information about physical–mechanical aspects of the world. The organs that detect this modality are referred to as **mechanosensory** organs. Different such modalities include gravity (position), acceleration (movement), changes in air pressure (sound), and mechanical pressure (touch). At the base of many of these senses we find the same simple multicellular structures: **statocysts** and **neuromasts** (Figure 12.3). Neuromasts are found only in vertebrates. The principle of both structures is the same. A small solid object, such as a crystalline particle (known as a **statolith**), is embedded within a liquid or gelatinous environment. Because the statolith is denser and has a larger inertial mass than its environment, it will move relative to the rest of the structure in response to directional acceleration. This movement activates sensory neurons in a directional fashion. In the simplest statocyst, the

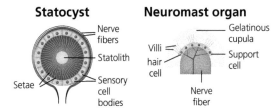

Figure 12.3 Mechanoreception organs: The structure of a statocyst (left) and a neuromast (right). Movement of a small but dense object (statolith) leads to the transfer of a neural impulse from the mechanoreceptor cell to the central nervous system.

statolith will move in response to gravity, indicating which direction is down. More complex and sensitive statocysts will be spatially arranged so as to give the animal information about acceleration in specific directions, thus providing feedback about the animal's own motion.

In vertebrates, neuromasts are at the base of a number of different sensory organs. Unlike statocysts, neuromasts also respond to the movement of liquid over them. Most aquatic vertebrates (but not marine mammals) have a **lateral line** system. This is composed of an elongated depression (or canal) in which are distributed numerous neuromasts. The flow of water within the canal differentially activates the neuromasts, giving information about the currents around the animal. The strength and direction of the currents can indicate the presence of other moving organisms, which could be predator or prey, and even signal when the animal is nearing an obstacle.

The lateral line system has evolved and been elaborated in vertebrates, including mammals, to give rise to two separate sensory systems that are both found in the inner ear: the **vestibular system** and the cochlear sound receptor. The vestibular system has the same function as the acceleration detecting statocysts described above, but a slightly different mode of operation. It is made up of three semicircular canals, all perpendicular to each other so as to cover three dimensions. Each canal is lined with neuromasts, just like the lateral line system. As the head moves or changes orientation, the liquid in the canals moves in response, providing feedback about the orientation and movement of the head. This system is very sensitive and is integrated with a feedback system that allows the body to correct

for even small movements, for example to keep the eyes fixed on a target as the head moves.

The **cochlea**, the ear's sound reception center, is an organ that detects and transmits information about vibrations in the air. The cochlea (Latin for snail shell) is a spiral organ that starts out thick at one end and tapers to a thin distal end. Air vibrations are picked up by the ear drum, a thin membrane covering the opening between the middle ear and the outside world. The vibrations are transferred via a group of three minuscule bones, the ear ossicles, to the end of the cochlea. The cochlea vibrates in response, but because of its structure, different regions of the cochlea vibrate in response to different sound frequencies. Low frequencies (low-pitched sounds) cause the thick end to vibrate and high frequencies (high-pitched sounds) cause the thin end to vibrate. Neuromasts along the cochlea thus respond to vibrations of specific frequencies. In essence, the cochlea performs what mathematicians call a Fourier transform—taking a complex repeating signal and breaking it down to its constituent frequencies. Information about the strength of different frequencies is then transferred to the brain. The remarkable process through which the separate sound frequencies are then reassembled by the brain to give us the subjective experience of hearing music or human speech are beyond the scope of this book.

Hearing is found outside of vertebrates, but nowhere is it as elaborate or as complex. Many animals have a hearing organ, not necessarily on the head, that vibrates in response to sound. This is usually tuned to a specific frequency or a few frequencies that allow them to pick up mating calls or aggressive calls of conspecifics. When the organs are paired, it allows localization of the direction from which the calls are coming.

Another type of sensory cell that detect physical aspects of the world is the touch receptors. These cells undergo deformation in response to pressure. Deformation of the cell causes it to produce action potentials. Mammalian touch receptors are found in the dermis (the mesodermal portion of the skin), and their distribution affects the sensitivity of different parts of the body. A high concentration of touch receptors in our fingers makes them very sensitive to tactile input, allowing us to differentiate different textures.

Cells with a similar function are found in other phyla, such as arthropods, in the form of sensory hairs. These are brittle structures that are connected to sensory cells that deform when the hair moves. As in the previous example, deformation of the sensory cell causes it to produce an action potential. Depending on the location of the hair cell, its size, and physical attributes, it can respond to the movement of air or to physical touch. Insect antennae are covered in minute hairs that respond to contact with the substrate or with obstacles. Many terrestrial arthropods have posterior structures covered in wind-sensitive hairs that alert them to the presence of potential predators coming from behind. It's worth noting that in arthropods, chemical and mechanical senses are often found on the same structure. An insect antenna is covered in both touch-sensitive hairs and chemical receptors.

The final physical sensory modality we will discuss is **proprioception** or self-sensing. Proprioception works through the action of mechanical receptors in muscles and joints that give the animal information about its position and activity. Through an integration of the input from all these receptors, the organism knows where all of its structures and appendages are. Proprioception is what allows us to touch our nose with our finger when our eyes are closed and allows us to walk and run without looking at our legs.

Light reception

Vision involves gathering photons from the environment. The basic unit of the vision system is the **photoreceptor cell**. This is a cell that has a membrane bound protein called opsin, which is linked to a pigment molecule or chromophore. Together, they have the unusual characteristic of undergoing a conformational change when hit by a photon. This change activates a series of downstream interactions, ultimately leading to the photoreceptor cell firing. Different combinations of opsins and chromophores are sensitive to photons of specific wavelengths.

The most primitive light receptors are clusters of photoreceptor cells that only collect light hitting them directly (Figure 12.4A). This type of sense organ provides information about the presence or absence of light, and possibly about its intensity.

If the organ is paired, comparing the signal from the two sides provides some information about the direction from which the light is coming. Throughout evolution, more complex light-receiving organs, or eyes, have evolved numerous times in different lineages. We can see a range of different eye types, each of which provides a certain type of visual information that is suitable for the organism. Because animals have different types of eyes, we can trace a hypothetical sequence of evolutionary events leading from primitive light receptors to the complex image-forming eyes of mammals (Figure 12.4).

After the appearance of a simple patch of pigmented cells, the next step would be the sinking of the middle of the patch to form a concave **cup-eye** (Figure 12.4B). The advantage of this development is that the information about the presence of light would also include where the light was coming from, since different parts of the eye would react to photons from different directions. Such eyes also detect changes in the amount of light over the eye over time, thus detecting motion. The larger and deeper this cup-eye, the more detailed spatial information it would provide. As the eye becomes more sensitive it can identify not only direct light, but also light that is reflected from other objects. The amount of light reflected by an object is a function, among other things, of its chemical makeup and physical texture. This is information that can be used by an animal to have a better understanding of its environment. Including opsins that respond to different wavelength provides information not only about the presence or absence of reflected light, but also about what object the light is being reflected from. As the cup-eye becomes larger and deeper, at some point it is transformed to a **camera eye** (Figure 12.4C), a slit-eye or a pinhole-eye. An eye that is composed of a photoreceptor-cell-lined space and is connected to the outside by a thin opening or hole can focus light on specific cells to form an image. This type of eye already provides detailed and nuanced information about the surroundings, and is found in some animal taxa, most famously in the nautilus—a relative of squid and octopus that retains many primitive features. The next step is the evolution of a simple lens and a **cornea**—a transparent covering to the eye opening, which provides additional resolution and clarity (Figure 12.4D). The final step is the appearance of a flexible lens under

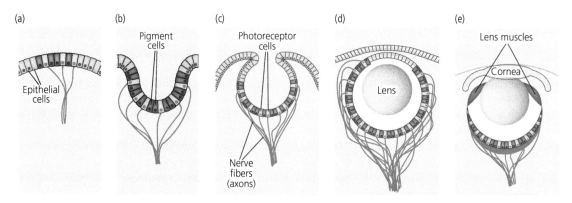

Figure 12.4 Types of photoreceptor organs: Different photoreceptor organs, arranged in a sequence of increasing complexity. All these types of organs are found in extant animals, and they represent a hypothetical sequence in which complex photoreceptor organs may have evolved. (a) Simple pigment patch with photoreceptive cells, which send nerve fibers to the central nervous system. (b) Cup eye, in which the photoreceptive cells are arranged in a concave structure. (c) Camera eye, in which the connection to the outside world is through a small opening, known as a pinhole. (d) Lens eye, in which a transparent layer covers the eye, and a focusing lens appears inside the eye cavity. (e) Complex eye, in which the transparent layer forms a distinct cornea, and the shape of the lens can be modified by muscles connected to its margins.

or within the cornea (Figure 12.4E). This type of eye allows focusing on specific objects, thus separating them from the background and allowing their identification. Adding more types of opsins allows the detection of different colors. This is the type of eye found in vertebrates, including ourselves.

Concomitantly with the elaboration of the eye's resolving power, the organism also has to evolve a nervous system that can interpret and process the complex information arriving from the eye. There is a correlation among animals between the complexity of the eye and the size and complexity of the brain.

Arthropods took a different evolutionary pathway and evolved a different mechanism for increasing the complexity of the eye. Rather than improving the sensitivity of the simple pigment patch, they repeat it multiple times, giving them compound eyes with numerous individual facets or **ommatidia**. It is unclear whether this process happened convergently a few times in different arthropod lineages or whether it dates back to very early in arthropod evolution.

Our perception of the visual world is filled with color. We are capable of seeing color because we have three types of opsins that are sensitive to red, green, and blue wavelengths. Our interpretation of colors is largely subjective. For example, we interpret pure green light in the same way that we interpret a mix of blue and yellow light. Animals with a different mix of opsins or a different processing system would not necessarily see these as the same color. Other animals have opsins tuned to different wavelengths and see the world very differently from the way we do. Many animals, bees and birds being well-known examples, see well into the ultraviolet part of the spectrum, which is invisible to us. Other animals, including some snakes, see infrared wavelengths. Different combinations of receptors can lead to the ability to differentiate colors that may seem identical to us. Many insects have five different opsins with different preferred wavelengths. There are reports of dragonflies with over a dozen different opsins. We can't even begin to imagine how the world of color is perceived by such an organism.

Other sensory modalities

Chemical, physical, and visual modalities are the main sources of information we have about the world, and they are the most common among animals. There are a few other modalities that are worth mentioning briefly for the sake of completeness.

Many animals are able to sense various types of magnetic and electromagnetic fields, not only radiation in the visible spectrum (light). It is clear that many animals are able to navigate using the Earth's magnetic field, which gives a constant signal indicating the direction of magnetic north. The precise mechanism for this sensitivity is unclear, although it is probably linked to the presence of magnetic materials in the animals' central nervous system. Some aquatic predators are able to sense electromagnetic fields that are generated by muscle activity in potential prey species. This is known as **electroception** or **electrolocation**. In vertebrates, the organs that detect electromagnetic fields are located in the front of the head and are known as the **ampullae of Lorenzini**.

Heat and temperature sensation are two separate sensory modalities. Sensing temperature is mostly a physiological mechanism that is not directly connected to sensory cells. Animals modulate their activity based on the external temperature or their internal temperature. Sensing heat or cold is connected to specific types of touch receptors. Vertebrates and other animals have an immediate withrdrawal response when touching very hot or very cold objects. Intriguingly, the neural processing system that responds to heat is linked to the chemical sense that detects capsain—the "hot" ingredient in chili peppers. Similarly, the system that responds to cold is linked to the system that responds to menthol, the "cool" ingredient in mint. Thus, our subjective feeling that peppers are hot and mint is cool are due to the neural connection between unrelated sensory modalities.

Pain, or **nociception**, is a separate sensory modality from touch receptors. It is mediated by distinct sensory cells and processed in a specific part of the central nervous system—at least in mammals. The evolutionary history and phylogenetic distribution of separate nociception is unclear.

An unusual sensory modality is the sensation of time. Time is not a tangible external stimulus like all the other examples we have covered, but the passage of time is important to an animal. Almost all animals have an internal clock that provides information about where (or when) in the cycles of

---

Box 12.1 The Cambrian Explosion

The first evidence of life in the fossil record dates back to just over 3.5 billion years ago. For the first three billion years of life's existence, it was fairly simple, consisting entirely of single-celled organisms. The complexity of animal life grew dramatically over a relatively brief period at the beginning of the Cambrian period about 540 million years ago. This dramatic increase in complexity has been dubbed the Cambrian Explosion.

Despite its name, the Cambrian Explosion was apparently not quite so explosive, but a rather long and drawn-out process, starting a few tens of millions of years before the base of the Cambrian and lasting into the middle of the period. Multicellular life first appeared during the Ediacaran period, about 570–540 million years ago, with a series of bizarre organisms, many of which cannot be easily connected to organisms we know today. Nonetheless, recent work has managed to link at least some of the Ediacaran biota to Metazoa, indicating that the roots of the Cambrian Explosion extend deeply.

The Ediacaran biota underwent slow changes until it mostly went extinct in the transition to the Cambrian. The Cambrian explosion of animal diversity did not arise out of nowhere, as it is often presented, but out of the embers of the Ediacaran. The first few million years of the Cambrian saw unusually rapid evolutionary processes, leading by the middle of the Cambrian, about 515 million years ago, to the representation of almost all known animal phyla in the fossil record.

The weird and wonderful animals of the Cambrian have been made famous by numerous representations in popular culture. They include such oddities as *Opabinia* with its five eyes and flexible proboscis, *Anomalocaris*, the giant predator of the Cambrian seas, *Hallucigenia*, a long and thin animal with spikes on its back, and the scale covered wormlike *Wiwaxia*. Extensive work on these and other Cambrian fossils has shown that as alien as they may seem, almost all of them can be accommodated well within the diversity of animals we know today, mostly as early sister taxa to familiar phyla. Understanding these ancient animals and how they fit in the Tree of Life has been crucial for understanding the early evolution of animal body plans. Studying the dynamics of the Explosion has taught us about how diversity evolves and has led to a better understanding of the how the world we live in came to be as it is.

time an organism is. The most ubiquitous are circadian clocks—internal clocks that are roughly one day long in duration and give information about the time of day. The basis of these clocks is a series of molecular interaction with a negative feedback loop that generates a molecular cycle of roughly 24 hours. The cells that generate the circadian clock are not sensory cells but usually cells within the central nervous system. These cells receive a synchronizing input, usually the beginning of daylight, from dedicated cells in the visual system, which keeps the circadian system in tune with the surrounding world. It is the mismatch between the circadian clock and the external world that leads to jetlag and similar phenomena. Some animals have internal clock that measure longer cycles, such as the lunar cycle, annual cycles. The details of how these longer cycles work are not clear, but in many cases they are crucial to many organismic functions, such as synchronizing reproduction, tracking seasonal prey items, or migrating with the changing seasons.

CHAPTER 13

Platyhelminthes

Introduction to Spiralia

Spiralia is one of the three main branches of Bilateria, together with Ecdysozoa and Deuterostomia, which we will come to later in this book. It was identified as a taxonomic group only in the late 1990s, following the advent of molecular phylogenetics. Analyses carried out at that time found that a number of phyla that were not previously considered to be related formed a strongly supported monophyletic group. These phyla were very diverse and there was initially no obvious character uniquely in common to all phyla in the group. The researchers who recovered this grouping noticed that some of the phyla are characterized by a typical feeding organ known as a lophophore (see Box 17.1), whereas others have a typical larval form known as a trochophore (see Chapter 10), and thus named the super-phylum Lophotrochozoa. It subsequently emerged that many the phyla in this grouping share a similarity in early development—spiral cleavage, which we will discuss in Chapter 25. Based on this similarity, most researchers and most texts have reverted to the older term Spiralia, which originally encompassed a somewhat different group of taxa. The newer taxon name Lophotrochozoa is used for a subset of phyla within Spiralia (see Box 13.1), although some workers see the two as equivalent. We follow the more recent convention and use Spiralia in the broader sense.

Spiralia includes nearly half of all recognized animal phyla, and includes animals as diverse as mollusks, segmented worms (annelids), moss animals (bryozoans) and flatworms, which are the subject of the rest of this chapter. The internal relationships among members of Spiralia are still under debate, and we shall not go into them here. However, several characters unite subgroups within Spiralia (such as the presence of trochophore larvae as mentioned above), and molecular phylogenies consistently and robustly recover Spiralia. We can therefore accept this to be a valid taxon despite the early controversies surrounding it.

The platyhelminth body plan

The first spiralian phylum we discuss is Platyhelminthes, commonly known as the flatworms. Platyhelminths have a body plan lacking a coelom (body cavity—see Chapter 16) and a through gut, and were historically considered to represent the ancestral body plan of Bilateria. Therefore, platyhelminths were placed at the base of Bilateria, as a sister group to all other bilaterian taxa, or in older classifications as ancestral to all Bilateria (see Box 16.1). One of the surprises of the molecular phylogenies from the late 1990s was the placement of platyhelminths within Spiralia. This affinity is now firmly established and indicates that the "simple" platyhelminth body plan is secondarily derived, and represents a specialized rather than primitive, condition. This intuitively surprising reduction in complexity is worth further discussion, and we will come back to it in Chapter19.

Although most phyla have lineages within them that have evolved parasitic lifestyles (see Chapter 14), Platyhelminthes is unusual in being dominated by species that live as obligate parasites and have diverged significantly from the typical body plan of the phylum. We will first describe the characteristics of the non-parasitic, or free-living,

Organismic Animal Biology. Ariel D. Chipman, Oxford University Press. © Ariel D. Chipman (2024). DOI: 10.1093/oso/9780192893581.003.0013

members of the phylum, and later elaborate on the main groups of the parasitic forms when we discuss platyhelminth diversity.

As their name suggests (both their common name and their scientific name), platyhelminths are flat worms (Figure 13.1). Recall that in Chapter 10 we

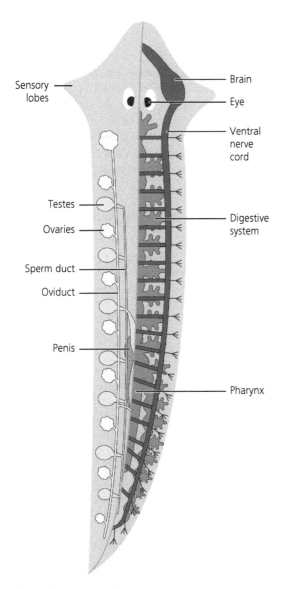

Sensory lobes

Brain

Eye

Ventral nerve cord

Testes

Ovaries

Sperm duct

Oviduct

Penis

Digestive system

Pharynx

Figure 13.1 Overview of the arrangement of major organ systems in free-living flatworms: A view from above of a typical free-living flatworm, with elements of the digestive system and the nervous system illustrated on the right half and elements of the reproductive system illustrated on the left half.

identified "worm" as a nonspecific term referring to a range of elongated bilaterally symmetrical animals with no external appendages. Platyhelminths are thus triploblastic and bilaterally symmetrical, with a clear anterior–posterior axis. However, unlike most bilaterians, they have a digestive system that has only one opening on the ventral side.

Most platyhelminths are small, rarely reaching more than a few centimeters in length, although significant exceptions can be seen in groups such as tapeworms. Their bodies are dorso-ventrally compressed, appearing as elongated ovals, flat arrows, or longer ribbons. Because of their flat shape, oxygen can reach all the cells by diffusion, and they have no circulatory system. In many taxa, the digestive system branches out throughout the body, providing direct nourishment to all tissues. There is no internal body cavity, and the internal organs are surrounded by a fairly loose population of **mesenchymal cells**.

Platyhelminths are also unique in their use of stem cells for all growth and cellular replacement even post-embryonically. Unlike most animals in which differentiated cells, like muscle cells, can divide and give rise to new muscles cells, in flatworms, differentiated cells do not divide and all new cells are derived from a population of undifferentiated stem cells that is maintained through adulthood. This unusual condition underlies the spectacular regenerative and proliferative abilities of many groups of flatworms as we will see below.

Platyhelminth diversity

The free-living flatworms include a large diversity of different lineages that have been historically grouped together in the class Turbellaria. However, they do not form a natural (i.e., monophyletic) group, as some lineages share a more recent common ancestor with the parasitic lineages of flatworms. Free-living flatworms include marine, freshwater and semi-terrestrial species, and indeed flatworms occupy effectively all environments that are not arid. The largest and best-known marine order is the polyclad flatworms or Polycladida. These are medium-sized to large worms with an oval body and often striking color patterns. They move by swimming or by gliding across surfaces usually via ciliary action, and are often found in littoral areas, tidal pools, and rocky shores. Another well-known group is the triclad flatworms (sometimes known as planaria), Tricladida, which include the familiar freshwater genus *Dugesia*, which is frequently used in teaching labs to demonstrate the platyhelminth body plan and animal regeneration. The triclad flatworms include several families that are semi-terrestrial, meaning they live outside the water, but rely on damp surroundings such as forest floors for survival. These semi-terrestrial platyhelminths have a highly ciliated ventral surface that allows them to glide across moist surfaces. The hammerhead worms (*Bipalium*) are an example of this lifestyle, with long worm-like bodies and a broad hammer-shaped head. They are voracious predators of snails, and several invasive species in North America and Europe, as well as some Pacific islands, present a significant threat to local snail populations. In addition to these familiar taxa, there are numerous smaller taxa, including many small and interstitial (i.e., living between grains of sand) species that are rarely seen by lay persons and less studied by professionals.

As stated previously, the majority of platyhelminths in terms of species numbers live as obligate parasites. However, unlike the free-living lineages, the parasitic lineages do form a monophyletic group called the Neodermata ("new skin") and include three main lineages: flukes (Trematoda), tapeworms (Cestoda), and monogeneans (Monogenea). As the name implies, neodermatans share the common feature of replacing their larval epithelium with a syncytial integument as adults. Parasitism therefore arose only once during the evolution of flatworms. Nonetheless, there is a propensity for symbiosis in the phylum, and there are many examples of free-living flatworms that have formed symbiotic relationships with other animals, some of which come close to parasitism (see Chapter 14).

Monogeneans are (mostly) external parasites of fishes and occasionally other vertebrates such as frogs. They have a direct life cycle involving live birth or egg-laying and are found on both marine and freshwater hosts. Most live on the gills or scales of fishes and can cause significant economic damage to some fisheries. They have a posterior attachment organ called an **opisthaptor** which contains suckers and hooks and can be highly elaborated, while the upper body has an anterior mouth and is highly mobile.

Unlike monogeneans, the flukes and tapeworms are both internal parasites of vertebrates and have highly complex life cycles involving multiple host species. In flukes, first larval stages almost always infect mollusks such as snails in which they undergo asexual multiplication, producing large numbers of a free-living stage called **cercariae**. These actively seek the next host, which is normally a vertebrate, and these second and sometimes third intermediate hosts are necessary to transfer the worms to the final, or definitive, host. Within the **definitive host** they transform into adult worms that live inside the body, typically the digestive system. Flukes are by far the most diverse group of flatworms (over 30,000 species.) They include the schistosomes, or bloodflukes, which are unusual in living in the circulatory system of the host. Bloodflukes are estimated to infect over 60 million people in the tropics.

Tapeworms are perhaps the most widely recognized type of flatworm, having a highly divergent adult body plan that is strobilar, or "segmented" (Figure 13.2). They also lack a gut completely, and instead absorb nutrients directly from their host across their modified integument. Like flukes, tapeworm larvae must develop in an invertebrate intermediate host (typically an arthropod), which is followed by one or more vertebrate hosts. Unlike flukes, all transmission between hosts is by ingestion, and is thus passive. As adults they live in the digestive system where they develop a long chain of segments, or proglottids, each containing a set of male and female reproductive organs. At the anterior end is the **scolex**, a club-like structure with hooks that serves to connect the tapeworm to the host's intestinal epithelium. The vast majority of their adult body plan is dedicated solely to reproduction. Like flukes, there are many species of tapeworms that are of medical importance to humans and of economic importance as they cause agricultural losses. The pork tapeworm, *Taenia solium*, is one of the best known examples and is the cause of one-third of epilepsy cases in Central America as result of human infection with tapeworm larvae.

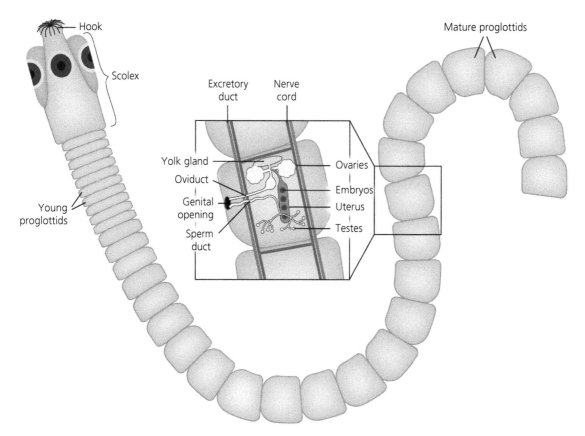

Figure 13.2 Body organization of cestodes: A schematic representation of a typical cestode tapeworm, with the internal structure of an individual proglottid enlarged.

Platyhelminth life history

All members of Platyhelminthes are **hermaphrodites**, having male and female gonads and reproductive organs. In most cases, fertilization is mutual, with both partners receiving and delivering sperm. Fertilization is usually internal and eggs are either laid in a protective foam covering or develop internally.

Most free-living flatworms lay a relatively small number of yolky eggs, investing a significant amount of energy in ensuring their survival. Yolk glands, which produce the yolk that is deposited in the forming egg, are a major constituent of the reproductive system, providing nourishment for the developing embryo. Eggs of free-living flatworms hatch as swimming larvae in some taxa and as miniature adults in others.

In addition to sexual reproduction, many platyhelminths reproduce asexually through fission or other means, and some can switch between the sexual or asexual reproduction depending on environmental conditions. In fission, the animal splits in two, with each half regenerating all the tissues in the missing half to give two fully formed offspring. Indeed, the regenerative capabilities of platyhelminths are highly developed. Most free-living lineages have high regeneration capacities, while the complex, proliferative life cycles of parasitic species can seem as a type of regeneration. Among all flatworms, planarians have the best regenerative abilities and can regenerate a full body from a fragment of tissue containing a single stem cell. They can even regenerate the central nervous system, the only animals known to have this ability. The extraordinary regenerative abilities

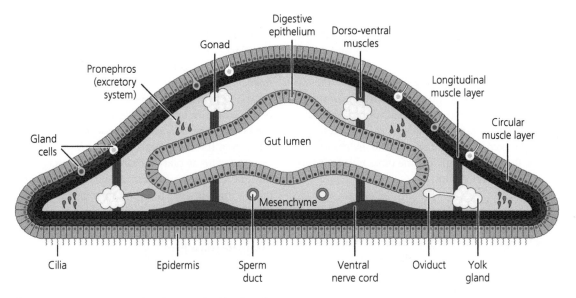

Figure 13.3 Transverse section through a generalized free-living flatworm.

of planarians have been recognized since the 1800s and they have been a used as a laboratory model for investigating regeneration for more than a century.

Parasitic flatworms have complex and specialized life cycles, often with several larval stages, each adapted to living in a different host. This is most extreme in trematodes, which can go through several intermediate hosts before settling into their main host. We will give some examples in Chapter 14, when we discuss parasitism. Like other platyhelminths, they are hermaphrodites, with well-developed male and female sexual organs. In cestodes, the sexes mature sequentially, allowing male and female proglottids of the same individual tapeworm to mate with each other. In most parasitic platyhelminths (and in parasites in general), the reproductive system is the most significant organ system, and often takes up a major proportion of the adult animals' body, with fully mature individuals being mostly an egg sack.

Feeding and digestion

Most free-living flatworms are carnivores, but some are also scavengers. They feed through an **eversible pharynx** that allows them to engulf their prey and start digesting their food externally. The pharynx is highly muscular, and the preliminary breakdown of large food items is done within the pharynx. Digestive glands that produce the enzymes and other constituents of digestive fluids lie beside the digestive system (Figure 13.2). Some flatworms have relatively simple, tube-like digestive systems, whereas others have highly branching intestines. Generally speaking, the larger the animal, the more complex the digestive system's branching pattern (Figure 13.3).

All other flatworms are parasitic, living inside or, more rarely, outside their hosts. As parasites, they have a ready supply of food that is often predigested. They therefore have either degenerate digestive systems or no digestive system at all, as is the case with cestodes.

Muscles and movement

Turbellarians move mostly by ciliary motion, gliding across the surface using their ventral ciliated epithelia. Larger members of the phylum can move through muscular movements, allowing them to swim in the water column. Indeed, some large marine turbellarians are very graceful swimmers. Muscles include dorso-ventral muscles and two muscle layers surrounding the body: a longitudinal layer and a circular layer. The coordinated activity of these muscles produces a rhythmic wave of motion of the lateral margins of the body and drives swimming.

In parasitic forms, larval movement mostly involves ciliary action although muscular control of hooks may be involved in penetration of the host. Adult worms, like free-living species, move through muscular action and may migrate externally or within the body using the same three muscle layers found in free-living forms.

Integument

Free-living platyhelminths have a multilayered epithelium as their body wall. Specialized glands and sensory nerve endings are embedded in this epithelium. The epithelium is ciliated, with a higher density of cilia on the ventral epithelium. Single cell glands secrete mucus for protection and as a substrate for locomotion or as a mechanism of attaching to the surface. Under the epithelium is a basement membrane, which in some taxa is thick and strong enough to act as a supportive external skeleton.

Parasitic forms have a larval epithelium that is similar to that of free-living species. However, when they mature, they shed the larval epithelium, and it is replaced by a specialized body covering, known as the **neodermis** as discussed above. The neodermis is a syncytial epidermis whose prime function is to protect against the **immune system** and/or the digestive system of the host. This specialized integument seems to have evolved once in the common ancestor of parasitic flatworms and is the main character uniting them as a monophyletic group known as Neodermata.

Nervous and sensory systems

The nervous system is composed of an anterior brain that is not very strongly differentiated from the rest of the nervous system. Posterior to the brain are two ventral nerve cords that include the neural cell bodies. These cords are often connected by lateral connections or **commissures**, forming a ladder-like nervous system. Parasitic flatworms typically have dorsoventrally paired nerve cords, running along the lateral margins of the animal.

The externally obvious sense organs are simple laterally facing anterior cup eyes that detect light and dark. Light-sensing organs can also be found in some larval cercariae of flukes, which live in open water environments. In addition, they have complex chemoreceptors, both along the body and concentrated in specialized organs on the sides of the head. These concentrations of sense organs are what give the arrowhead appearance of the anterior part of many platyhelminths. Statocysts providing gravitational information and touch receptors on the integument are found in some, but not all, platyhelminths.

Evolutionary history

Platyhelminths are almost unknown in the fossil record, being small, soft bodied, and frequently living as internal parasites. Their fossil record mostly includes the hard spines of parasitic forms and the remains of parasite eggs, often found in fossilized feces (**coprolites**) or even within fossils of the hosts. These fossils indicate that parasitic platyhelminths have been around since at least the Devonian (about 400 million years ago). We can also learn about the evolutionary history of parasites from the evolution of their hosts, since parasites share a co-evolutionary history with their hosts. This line of reasoning suggests that Neodermata probably evolved with bony fish about 450 million years ago. Fossils of free-living platyhelminths are only known from amber deposits in the Cenozoic, about 50 million years ago, but since Neodermata evolved from Turbellaria, we can be fairly certain that their evolutionary history extends much further back.

Since parasitic forms evolved from free-living ones, and free-living platyhelminths are themselves missing many characters relative to ancestral spiralians, we must assume a fairly rapid evolution of platyhelminths shortly after the split between Spiralia and other bilaterians. This probably happened at least 500 million years ago. The two main unusual characters of platyhelminths are the loss of the through gut and of the coelom, and the use of somatic stem cells for differentiation (Figure 13.4). An interesting question is what selective pressures led to these two characters. While there is no clear answer to this question, it may have to do with a

(a)   (b)   (c)

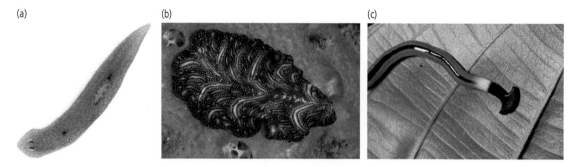

Figure 13.4 Examples of platyhelminth diversity: (a) A freshwater tricladid turbellarian, *Cura foremanii*. (b) A polyclad turbellarian, the Persian carpet flatworm *Pseudbiceros bedfordi*. (c) A terrestrial tricladid turbellarian, the hammerhead worm *Bipalium*.

Source: a: Photo supplied from Shutterstock: Ernie Cooper (https://www.shutterstock.com/image-photo/specimen-freshwater-tricladid-flatworm-planarian-cura-2005284503); b: Photo supplied from Shutterstock: Jack Pokoj (https://www.shutterstock.com/image-photo/persian-carpet-flatworm-1267827529); c: Photo supplied from Shutterstock: Pong Wira (https://www.shutterstock.com/image-photo/hammerhead-flatworm-landplanarian-bipalium-on-726631237).

shift in life history that led to their becoming small and interstitial. We will come back to this question in Chapter 19, where we will see that secondary simplification is often linked to a reduction in body size.

Box 13.1 Gnathifera

The exact relationships among the many lesser-known spiralian phyla have been debated extensively since Spiralia was first identified as a distinct clade. This is largely due to the minimal representation of these phyla in most analyses that use modern approaches. In this Box, we will briefly survey a subset of these lesser-known phyla, which is usually considered to represent a sister group to all other spiralian phyla. Most of the phyla within this Box belong to a clade that is known as Gnathifera, characterized by a special type of cuticularized jaws (the literal meaning of the taxon name is "jaw-bearing"). The term Lophotrochozoa is now used by many researchers to refer to the clade that includes the remainder of spiralian phyla not included within Gnathifera.

Rotifera—the wheel animals. If you look at a sample of water from a puddle or pond you are almost sure to find rotifers. Indeed, for a "lesser-known" phylum, they are extremely abundant organisms. About 2000 species of rotifers are known. Rotifers are mostly microscopic, ranging in size from a few tens of microns to 1 or 2 millimeters. In a typical freshwater sample they may be smaller than some of the single-celled ciliates found with them. Rotifers bear a pair of anterior ciliary organs that look superficially like rotating wheels, giving both their

common and scientific names. One branch of Rotifera, the bdelloid rotifers, presents a fascinating evolutionary paradox, since the species within the branch are composed entirely of females and have been reproducing asexually for millions of years.

Acanthocephala—literally "spiny-heads." This is a group of about 1200 intestinal parasites of vertebrates that can reach tens of centimeters in length. Most recent phylogenies place them as a group within Rotifera.

Chaetognatha—arrow worms or "spiny jaws." This is another example of an extremely abundant "lesser-known" phylum. The phylogenetic position of chaetognaths has been notoriously difficult to ascertain, and over the years they have been placed in almost every conceivable position within the animal Tree of Life. Today they are generally accepted to be either within Gnathifera or as sister to all other spiralians. Chaetognaths are planktonic marine worms, reaching from a few millimeters up to a few centimeters in length, with a typical arrow-shaped body and a mouth surrounded by spines or teeth adapted to capturing prey. While there are only about 200 known species, in terms of biomass, they make up a significant portion of the zooplankton, possibly second only to copepods (see Chapter 23). They are likely the most significant predators in the plankton.

Two other phyla are worthy of at least a mention. These are Gnathosomulida and Micrognathozoa. Both are composed of minute worm-like animals that live on or within the sediment. Both include only a handful of species. Micrognathozoa has the distinction of being the newest animal phylum, only being officially described in the year 2000.

CHAPTER 14

Parasitism

What is a parasite?

Parasitism is a mode of life in which one organism, the **parasite**, lives on or in another organism, the **host**, in a relationship that benefits the parasite while causing harm to the host. Parasitism can be seen as a type of long-term symbiosis, where the parasite is dependent on the host, but the host is better off without the parasite. In many cases, a parasite feeds on the host's tissues themselves, or on fluids it contains (e.g., blood-sucking parasites). In other cases, it takes advantage of metabolites or partially digested food produced by the parasitized organism (e.g., intestinal parasites). The host not only provides the parasite with a source of food, but often, unwittingly provides the parasite with protection and in many cases forms the parasite's entire environment.

Parasitic organisms are often functionally divided into **ectoparasites** and **endoparasites**. Ectoparasites live outside the organism, usually on or in the integument. They can attach temporarily and detach when they have finished feeding (this is true for many blood-sucking parasites), or they can spend their entire life or the entirety of one stage of their life cycle attached to the host once they have located and successfully occupied it (e.g., lice or fleas).

Endoparasites live inside their host. They can live inside tissues, such as muscles or digestive glands, or they can live inside body cavities, and especially within the digestive tract of the host, as for example, the tapeworms and flukes we met in the previous chapter. Endoparasites tend to be extremely specialized, having evolved strategies to invade the host, locate the target tissue, and move toward it, all while evading the host's immune system.

Parasitism is a highly successful evolutionary strategy. Once the parasite has successfully occupied its host, it does not need to invest energy in finding resources or in protecting itself from the external environment (in the case of parasites that permanently live on or in their host). Parasitism has therefore evolved numerous times in animals, with almost every phylum having at least some parasitic species, and some phyla such as nematodes (Chapter 20) or platyhelminths (Chapter 13) being predominantly composed of parasitic species. Parasites are common outside of the animal kingdom, with many known examples of parasitic fungi, unicellular eukaryotes, and even plants that parasitize other plants.

Parasites are often species specific, and every animal species has its own parasites. This statement is probably also true for the parasites themselves, who may have their own parasites. We can conclude from this that parasitic animal species probably outnumber free-living species by an order of magnitude.

The host–parasite arms race

By definition, parasites cause some degree of harm or even **morbidity** in the host. A well-adapted parasite does not kill its host, at least not immediately, since it relies on the host for its nourishment, protection, and ultimate survival, and excessive harm leads to an evolutionary dead-end for the parasite. However, the parasite usually does reduce the host's fitness, by weakening it and reducing its chances of finding food and avoiding predators and, more importantly, reducing its chances of mating and reproducing. There is therefore a strong selective advantage to evolving improved defenses

Organismic Animal Biology. Ariel D. Chipman, Oxford University Press. © Ariel D. Chipman (2024). DOI: 10.1093/oso/9780192893581.003.0014

against parasites, either through modifications to the immune system or the tissues that are targeted by the parasite, or through behavioral changes that reduce the chance of infection or remove parasites that have already infected the organism.

The parasites, in turn, evolve strategies to avoid the host's defenses, by changing the way they invade or modulate the host's defenses or by masking themselves from the host's immune system. The interaction between parasites and hosts leads to an accelerated **evolutionary arms race**, where both parasite and host continuously adapt to the changes taking place in the other. Consequently, parasites tend to have high rates of evolution, as seen both in their highly divergent morphologies and in the rate of molecular changes they undergo compared with non-parasitic relatives. This arms race is probably also what lies at the base of the high diversity of parasites and their specificity. Once a parasite evolves to avoid the defenses of a specific host species, it is likely to be not as well-adapted to infecting other species.

There are cases when a parasite does lead directly to the death of its host. In these cases, it is known as a **parasitoid**. Well-known examples are parasitoid wasps, which lay their eggs inside the host. The wasp larvae hatch and consume the host from inside, with the host ultimately becoming a dead shell in which the larvae grow. Parasitoids evolve complex methods to keep the host alive as long as possible so that it continues to provide nourishment. The host mostly evolves strategies to avoid being detected by the parasitoid, since once it has been infected, there is little it can do in terms of ridding itself of the parasite.

Complex life cycles in parasites

A common theme among parasites is that they have complex life cycles. Many parasites go through a number of life history stages, often switching between sexual and asexual reproduction at different stages. In many cases, this complex life cycle is also linked with a transition between different hosts, with each stage being adapted to a specific host. Some stages may be highly species-specific, as described above, whereas other stages are more flexible with regard to the host. The host in which the parasite carries out sexual reproduction is known as

the **primary host** (or **definitive host**) whereas hosts in which reproduction is asexual, or in which the parasite does not reproduce are known as **intermediate hosts**.

Because of the small chance for each stage to successfully invade its host, survive in it, and move on to the next host, parasites produce very high numbers of offspring to increase the chance of at least some of the offspring completing the complex life cycle. In many cases, a significant proportion of the parasite's body and a significant amount of its energy are devoted to reproduction, as we saw in parasitic platyhelminths in Chapter 13. Conversely, parasitic organisms do not invest in maternal care and thus rely on high numbers of offspring for the survival of the species. Beyond that, parasites have evolved some incredible strategies to increase their chance of success, including manipulating the host's behavior to enhance transmission of the parasite.

We will demonstrate complex parasitic life cycles with two examples from the platyhelminths we met in the Chapter 13. Similar examples can be found in many other parasitic taxa, but the exact details are as diverse as the parasites themselves.

Dicrocoelium is a liver fluke (Trematoda) of grazing livestock, mostly infecting sheep and cattle. The *Dicrocoelium* life cycle starts with eggs that are shed in the pasture. The eggs are picked up and eaten by land snails, the first intermediate host, where they develop into larvae known as a **miracidia**. The miracidia penetrate the gut wall and settle in the connective tissue surrounding the gut. Here they undergo asexual reproduction to produce additional larvae known as **cercariae**. The cercariae migrate to the snail's respiratory system, probably as a consequence of the snail's immune system attempting to rid itself of the parasite, and are released in slime balls, each containing several hundred cercariae. Ants typically follow snail slime trails for the moisture they contain. When an ant following a moisture trail and comes across a slime ball, it ingests it, thus becoming the next intermediate host for the fluke. In the ant, the cercariae form cysts (now called **metacercariae**) and remain in various tissues of the ant, but one or more of the cercariae migrate to the ant's brain, where its presence modifies the ant's behavior. A parasitized ant climbs up blades of grass and attach itself to the highest point, locking its mandibles to the leaf.

It will do this every evening, and stay attached until the next morning, behaving normally for the duration of the day. This position at the blade tips enhances that chances that the ants, together with their internal parasites, will be eaten by grazing animals. Once in the final host, the metacercariae exit the cysts in the digestive tract and mature into adult flukes. The adults migrate to the bile duct where they undergo sexual reproduction, producing eggs that are shed with the feces into the field, allowing the cycle to repeat.

The tapeworm *Taenia* has a less complex life cycle involving only two hosts. Eggs shed in the field are eaten incidentally by a herbivore, with different species of *Taenia* specializing in different species of herbivores. The eggs hatch in the intermediate herbivore host's digestive tract, penetrate the intestinal wall, and enter the circulatory system. The larvae then travel through the circulatory system to the target tissue, usually muscles and the nervous system. There they metamorphose into encysted larval worms. When infected tissues are consumed by the final host, the larvae exit the cysts in the stomach and establish themselves in the small intestine where they develop a body consisting of repeated segments, each containing male and female reproductive organs. The cycle is completed when the eggs or entire segments detach in the intestine and are shed into the environment with the feces. The case of the pork tapeworm (*T. solium*) is particularly unusual in that humans can act in either the role of the final or intermediate host. Thus, if infected and undercooked pork is consumed, the adult worm will develop in the intestines and can cause discomfort, but is readily treated through oral medication. However, if the eggs of the pork tapeworm are consumed by humans, then these will metamorphose into larvae that migrate to the neuromuscular system, including the central nervous system. Infection with larval tapeworms—that is, playing the role of intermediate host—thus results in more serious, and usually chronic, disease.

Secondary simplification and extreme specialization

A common characteristic of parasites is that they often undergo **secondary simplification**. As part of the arms race with the host, there is strong selection to improve characters that aid in host evasion and in reproduction, whereas there is no selection to maintain characters or systems that are not as important once the parasite is living within or upon the host. Consequently, many of these characters are lost over evolutionary time. Many parasites thus show significant simplification of sensory and nervous systems in the adult stage, they often lose structures related to motility (e.g., loss of wings in fleas and lice), and, in some cases, they lose parts or even the entirety of the digestive system, since they receive nutrition with little effort from their host (as seen in tapeworms). Many genes and proteins are also lost as part of this simplification of unnecessary systems.

An even more extreme example of simplification is seen in Myxozoa—cnidarian parasites that have degenerated to the extent that they do not undergo embryonic development, but reproduce via spores. Myxozoans never have epithelia, nervous systems, or digestive systems, all of which are structures found in all other cnidarians.

The combination of their rapid evolution and secondary simplification have historically made the affinities of parasitic lineages difficult to discern, and many were therefore grouped as independent phyla. For example, the Myxozoa are so radically different from other cnidarians that they were only recently discovered to be members of that phylum. A group of parasites known as pentastomids, or tongue worms, were long considered to be a phylum on their own, until molecular phylogenetics identified them as being highly derived crustaceans. Similarly, acanthocephalans, or thorny-headed worms, were a separate phylum prior to molecular phylogenetics convincingly showing them to be a parasitic lineage of Rotifera. Even when the higher-level taxonomy of parasites can be identified, the details of their phylogenetic position can be difficult to verify because their rapid evolution is also manifested in their rapidly evolving genomes, introducing complications and noise into molecular phylogenetics.

At the same time, the need to adapt to the host also poses high selection pressures for specialization. Thus, although parasites may lose significant parts of their biology that are no longer required in a parasitic lifestyle, making them appear "simple," they simultaneously show extreme adaptations relating to aspects of their parasitic lifestyles.

Concluding notes

Parasitism is one of the most fascinating areas of study within evolutionary biology. We are drawn to parasites with a mix of horror and amazement. Their intricate strategies for invading and infecting their host invite comparison with evil monsters of myth and fantasy. But closer to reality, parasitism is simply another feeding mode, like herbivory or detritivory. Other living organisms are a resource that can be exploited by parasitic species, just like any other element of the environment. Indeed, parasitism is likely the most common form of living on the planet and this is reflected in the extraordinary diversity of animals that have evolved to live parasitic lifestyles. Beyond their importance for evolutionary biology, parasites are an important component of the ecosystem, with a high proportion of animals being parasitized at some level for most of their lives. From the human perspective, parasites have been an important fact of life throughout most of our history and still are in many developing countries, either as a health risk or as a risk to livestock (see Box 14.1). It is important to remember that parasites are an extremely significant part of the biosphere, even if they are often overlooked.

Box 14.1 Major human parasites

No discussion of parasitism is complete without a mention of some of the main parasite species that affect humans. There are of course numerous detailed texts on human parasitology, and this Box is meant mainly as a preliminary pointer for further exploration in more detailed texts. Here, we will mention a handful of species, giving their phylogenetic position and an overview of their mode of parasitism. We restrict the discussion to metazoan parasites, ignoring the many examples of unicellular parasitic species, as well as fungi and bacteria.

Schistosoma is a genus of trematodes, commonly known as blood flukes. Several different species are known to infect humans, each being found in different parts of the world and each affecting different organ systems. As for most trematodes, freshwater snails act as intermediate hosts, and infection is most common in bodies of fresh water with snail populations. The disease caused by species of *Schistosoma* is known as schistosomiasis, bilharzia, or Katayama fever. It can appear both as an acute infection, manifested as a fever, and as a chronic infection usually affecting the digestive system or bladder, but in some cases spreading to other organs as well. It is estimated that hundreds of millions of people are infected with schistosomiasis, mostly in Africa, but also in rural communities in other developing countries.

Fasciola is also a genus of trematodes, commonly known as liver flukes. They have a complex life cycle, including a stage in freshwater snails. *Fasciola* can affect many species of mammals, including domestic animals and humans. The adult worms are fairly large and settle in the liver. Because of their size, they can damage liver function or lead to obstruction to bile ducts (the ducts that leads enzymes from the liver to the digestive system).

Several species of tapeworms (Cestoda) affect humans. We already mentioned the pork tapeworm (*Taenia solium*) in the main text. Humans can also by infected by the fox tapeworm (*Echinococcus multilocularis*), in which humans act as intermediate hosts. The primary hosts are mammalian carnivores, such as foxes or lions. Symptoms include the appearance of cysts in the liver, the lungs, or other organs.

There are several species of human parasites from the phylum Nematoda—roundworms (see Chapter 20). Filariasis is caused by nematodes of the family Filaroidea, which are transmitted by biting insects. The filarial nematodes can accumulate under the skin causing rashes in the eyes, leading to river blindness, or in lymphatic tissues, where their accumulation causes the swelling of tissues known as elephantiasis.

The nematode genus *Ascaris* consists of intestinal parasites that are unusually large for nematodes (up to 20cm in length). Infections by *Ascaris* are common worldwide, and *Ascaris* eggs have been found in archeological excavations of ancient sewer systems and in the intestines of Egyptian mummies. In extreme cases, infection by *Ascaris* can lead to life-threatening intestinal blockage.

Hookworms and whipworms are two additional common nematode parasite genera, each consisting of several species. Whipworms infect the digestive system, whereas hookworms are blood feeders. Both are extremely common and can be picked up from contaminated soil. While they constitute a health concern, neither is commonly lethal.

CHAPTER 15

Mollusca

An introduction to the mollusks

With close to 200,000 known species, Mollusca is the second largest phylum in the animal kingdom. It includes many familiar groups, found both on land and in the water. In discussing Mollusca in this chapter, we are moving on to more complex body plans than those we have met so far, and we will introduce many new concepts and organizational principles. The name Mollusca comes from the Latin word for soft, *mollis*, due to their soft body and lack of an internal skeleton. Note that the UK spelling is mollusc, the US spelling is mollusk, while the scientific name of the phylum is Mollusca.

The mollusks are highly diverse, with body plans ranging from the worm-like aplacophorans, to the sessile bivalves (clams, oysters, and their relatives), to the active cephalopods (octopuses and squid), with their complex behavior. They range in size from tiny millimeter-scale gastropods (snails) and bivalves to giant clams over a meter in length, and squid and octopuses with arms that extend to several meters. Nonetheless, we can reconstruct a generalized mollusk (Figure 15.1) and see how the diverse mollusk body plans are variations on a theme. Different lineages evolved different specializations, elaborating or reducing different elements of the ancestral body plan. Some lineages evolved toward increased activity and predatory behavior while some evolved toward sessile suspension feeding. Many lineages exhibit intermediates states between these two extremes. Several lesser-known taxa that have been overlooked for many years are now providing insights into the early evolution of the group and are bringing us closer to reconstructing the ancestral mollusk body plan, or at least the

ancestral body plans for the diverse lineages within Mollusca.

Mollusca is a predominantly marine phylum, and most of its evolution took place in the marine realm. Only one significant group of mollusks has left the sea, the familiar land snails. A few members of other lineages exhibit semi-terrestrial adaptations. Within the land snails there has been a secondary return to fresh water, and the freshwater snails still maintain many characteristics of their terrestrial history.

Molluskan body organization

Mollusks are bilaterally symmetrical, triploblastic animals with a secondary body cavity known as a coelom. We will discuss coeloms in general, their characteristics, and their evolution in Chapter 16. We will start by describing the general features of the mollusk body plan, and then see how this this body plan has been modified in the various taxa. The current understanding of mollusk evolution sees a deep split between a taxon known as Conchifera, which includes most of the familiar molluscan groups, and a taxon known as Aculifera, which includes several of the lesser-known taxa. The following description is largely based on conchiferans. Specific aculiferan characteristics are discussed where relevant.

The first obvious feature of mollusks is the fact that most have a calcareous **shell** of some kind. While many taxa have secondarily lost the shell or internalized it, it remains one of the defining characters of Conchifera (literally "shell bearer"). The shell is secreted by a specific epithelial organ known as the **mantle**. Even taxa that have lost the shell

Organismic Animal Biology. Ariel D. Chipman, Oxford University Press. © Ariel D. Chipman (2024). DOI: 10.1093/oso/9780192893581.003.0015

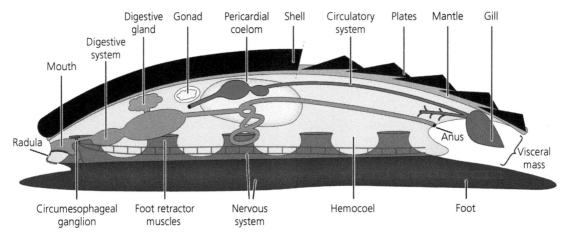

Figure 15.1 Main elements of the molluskan body plan: Lateral view of the main organ systems in a generalized mollusk. This illustration combines elements from different mollusk groups. Note the typical conchiferan shell in the anterior half and the typical aculiferan plates in the posterior half.

maintain the mantle. The shell is built of an organized crystalline structure that includes calcium carbonate as well as various organic components. In contrast, members of Aculifera do not have a single shell. Some have no shell (the aplacophorans), whereas others have seven distinct shell plates (the polyplacophorans). Based on the fossil record and on developmental data, seven plates are probably ancestral for Aculifera, with aplacophorans representing a secondary loss of the plates.

The mantle folds to form the **mantle cavity**. This is a space that is continuous with the outside world, although in some cases it can be closed by a dedicated muscle. Various organ systems open into the mantle cavity. The anus opens into the mantle cavity as do the excretory ducts. Sperm and ova are also released into this cavity in many cases. The mantle cavity is usually the site of gas exchange in larger species and it often houses a set of comb-like gills called **ctenidia**. This cavity is also the space a snail retreats into when it hides in its shell.

The main part of the mollusk body is composed of two structures: the ventral **foot** and the dorsal **visceral mass**. The foot is a muscular structure and is usually the main organ used for movement. It often has a ventral ciliated epithelium that adheres to the surface and provides some or all of the movement along the substrate. Most internal systems are housed within the visceral mass, including the digestive system, the circulatory system, and the gonads.

Molluskan life history

Many different life-history modes can be found in different molluskan taxa, usually depending on the level of motility. Mollusks are ancestrally indirect developers, but many taxa have lost the larval stage and hatch as miniature adults. The typical molluskan larva is a trochophore larva, with some marine snails having a veliger larva at a later stage (see Chapter 10 for details on these larval types.)

Mollusks normally live individually, although some species of sessile bivalves and snails form dense colonies. These colonies are never clonal. Rather, they are colonies formed by aggregation of individuals in a suitable habitat, with no division of labor or resource sharing among colony members.

Almost all mollusks reproduce sexually, and almost all variations on sexual reproduction can be found. Most mollusks have separate sexes, but many species are hermaphrodites, especially within terrestrial snails, with some rare cases of species that can self-fertilize. Reproduction is external in sessile marine species, with synchronized release of sperm and ova into the water. Semi-internal fertilization is found in some taxa, where ova are released into the mantle cavity and sperm cells from another individual enter and fertilize them there. Land snails and many aquatic snails have internal fertilization. In hermaphrodite species, copulation can either be reciprocal and simultaneous, or one member of the pair acts as a male and one as a female. Finally, in some taxa, notably in cephalopods, the male produces a special packaging for the sperm, known as a **spermatophore**, which he then passes to the female, who extracts the sperm from it and fertilizes her ova.

Feeding and digestive system

The molluskan digestive system follows the generalizations we outlined in Chapter 7, with specific adaptation depending on the feeding mode. The mouth leads to an esophagus that connects to a stomach with attached digestive glands. Digested food from the stomach moves to a coiled intestine for absorption. Aculiferans have uncoiled digestive systems. Digestive waste products leave the body via the anus, which empties into the mantle cavity. Feeding modes in mollusks are diverse, and almost all possible modes can be found in some taxon. Suspension feeding is the norm in bivalves, with water being passed through the shells where food particles are picked out and transferred to the mouth. Cephalopods are carnivores, with different species specializing in active hunting or in collecting sessile animals. Many of the lesser-known taxa are detritivores, scavenging as they move along the ocean floor, and this is probably the ancestral mode for Mollusca. Within the snails, nearly all feeding modes can be found—herbivory, predation, detritivory, and even filter-feeding in some sessile groups.

The **radula** is a unique molluskan feeding organ. This structure consists of chitinous teeth and is found inside the mouth. The radula is a scraping organ, used to collect food from the ocean floor or from rock surfaces. It is used for scraping in many detritivores and herbivores; however, it is modified within taxa with other feeding modes. In some carnivorous snails, it is modified to a spike or spear, used to drill into the shells of prey species or inject them with venom. Bivalves have lost the radula, since it is not needed for suspension feeding, and use modified ctenidia as their main feeding organ.

Circulatory system

This is the first group we have met to have a circulatory system (see Chapter 24 for a more detailed discussion of transport systems). The primitive type of molluskan circulatory system is an **open circulatory**

system. In the open system found in most mollusks, **hemolymph** (the circulatory fluid) flows through a main dorsal vessel that is muscular and pumps the hemolymph anteriorly. The hemolymph empties into a body cavity known as a **hemocoel**, where it bathes the various organs, providing oxygen and nutrients. It is then taken up from the hemocoel by the main vessel to complete the circulation. In cephalopods, uniquely, there is a **closed circulatory system**. In this system there is no hemocoel, and the blood flows through vessels that reach the entire body. Cephalopods have three muscular expansions of the blood vessel, or **hearts**, that drive the blood through the circulatory system. A main heart pumps blood through the body and two accessory **branchial hearts** drive blood to the gills. The heart sits within a cavity known as the **pericardial coelom**.

Gas exchange is usually through the ctenidia, which are filled with blood and absorb oxygen from the surroundings, regardless of the type of circulatory system. Many mollusks have lost the ctenidia and absorb oxygen solely through the body surface, or have secondarily evolved external gill-like structures. In terrestrial snails, the gas exchange takes place within the air-filled mantle cavity, and the cavity functions as lungs, with gas exchange taking place from the air within the lungs.

Muscles and movement

Ancestrally, the muscular foot is the main organ for benthic locomotion and is shaped like a flat sole. In some mollusks, foot retractor muscles extend into the visceral mass. Ancestral mollusks probably moved across the sea floor through muscular contractions of the foot, aided by the ventral cilia and by mucus produced by gland cells in the ventral surface. This mode of locomotion is still found in many benthic species, as well as in terrestrial snails and in polyplacophorans.

Some "worm-like" mollusks burrow into the substrate using **peristaltic** movement (see Chapter 16). Scaphopods and many bivalves also burrow using their foot. Some unshelled marine snails swim, often very gracefully, using synchronized contractions of muscles in the body wall. The most active mollusks are the cephalopods. The foot in cephalopods is modified into a funnel. The mantle cavity has become a jet-propulsion system, which can be filled with water that is then expelled at high speed through the funnel to push the animal rapidly in the opposite direction. In addition, cephalopods have arms or tentacles that are used for swimming in the water column or for moving along the substrate in a type of walking.

Many molluskan taxa have evolved a sedentary or sessile lifestyle. This is most obvious in bivalves, which live buried in soft sediment or firmly attached to hard surfaces. The foot in bivalves is modified either as a burrowing organ or as an adhesive organ that secretes strings of a sticky material known as **byssus**. Some bivalves have evolved a novel mode of locomotion, known as **valvular swimming**. In these species, the two valves open and close rapidly to push the animal backward through the water column. Some snails have also evolved a sessile lifestyle. In these, the shell is modified into a tube, with the tube usually cemented to a hard surface, either individually or in dense colonies. Sedentary snails, polyplacophorans, and monoplacophorans have a foot that is modified to cling to hard surfaces. These sedentary species forage in their close surroundings and often return to a single spot. In many cases, their shell grows so that it fits perfectly into their preferred spot, providing them with extra protection.

Integument

The molluskan body is covered by an epithelium, usually mono-layered and ciliated, which secretes a thin and soft cuticle. The cuticle is not very resistant to either water or gas entering. The epithelium is modified for shell secretion in the mantle region and as a dense ciliary epithelium in the ventral foot in many mollusks. The cuticle may include spines or spicules, especially in unshelled species. Many mollusks secrete mucus through the epithelium, either as an aid to locomotion, as detailed above, as a way to protect against dehydration in land snails, or occasionally, coupled with noxious chemicals, as a predator deterrent.

One of the main differences between Conchifera and Aculifera is in the type of external armor secreted by the mantle. In conchiferans it is a shell, usually made of a single unit, except in bivalves, where it is split in two. In aculiferans it is a combination of plates and spicules. Plates are found in polyplacophorans and in some fossil aculiferans and are individual articulating units—usually seven or eight in total. Spicules are small spines, found in aplacophorans. It has been suggested that the conchiferan shell as well as the plates of polyplacophorans are derived from ancestral spicules.

The cephalopod integument includes a diversity of pigment cells, which can change color, enlarge, or shrink to allow the animal to change color and patterns rapidly under neural control. Some cephalopods can also modify the texture of their integument to aid in camouflage against complex backgrounds.

Nervous system

The molluskan nervous system is tetraneurous in nature, meaning it consists of two pairs of longitudinal nerve cords that emerge from a more or less concentrated (ganglionated) brain. One (usually the medial) pair lies more ventrally to the other that is in turn situated more posteriorly and laterally. These nerve cords are usually interconnected with commissures. The ventral pair innervates the foot and the lateral pair innervates the visceral mass. The anterior of the conchiferan nervous system includes an expanded ganglion that encircles the anterior digestive system and is therefore known as the **circumesophageal ganglion**. This ganglion functions as the molluskan brain. In cephalopods, the brain is greatly elaborated and includes several ganglia, to give the most complex known central nervous system outside of vertebrates. Aculiferans have no real brain, with only a series of thickened commissures in the anterior.

Mollusks have highly developed and diverse sense organs. Eyes are found in many taxa but are missing from many others. When present, eyes are usually of the pigment-cup type, and are sometimes held on stalks. Cephalopod eyes are among the most complex eyes known, with variable focus lenses and high-resolution retinas. These eyes are very similar to vertebrate eyes but have evolved convergently. While most bivalves have lost their eyes, some groups have unpaired eyes along the margins of the shell. Polyplacophorans have numerous eyes along the shell, with lenses made of crystalline calcium carbonate.

Chemoreception is largely through specialized ectodermal sensory organs known as **osphradia**. Mollusks use chemoreception to find food, and gather information about their environment, but also to communicate, as chemical messages are often left in the mucus trails of snails. In bivalves, the chemosensory organs lie along the margins of the valves, which is the region most exposed to water. Some snails have specialized tentacles with a rich array of chemosensory cells.

The main mechanical sensory organs in mollusks are statocysts, which are found in almost all taxa. In snails, statocysts are concentrated in the muscular foot. Statocysts are highly developed in the active cephalopods and are complemented by complex tactile and propriosensory receptors along their flexible arms, allowing them to manipulate objects and probe their surroundings.

Molluscan diversity and taxonomy

Mollusca is traditionally divided into at least seven class-level taxa, although some sources divide them differently (See Figure 15.2). The first two taxa in the list below comprise Aculifera, while the remainder are members of Conchifera.

Aplacophora is a taxon of shell-less worm-like mollusks. The name means "bearing no shell." Aplacophora is composed of two taxa, which are sometimes considered to be individual classes, Caudofoveata and Solenogastres. These are benthic animals, some of which (members of Caudofoveata)

(a)
(b)
(c)
(d)

Figure 15.2 Some examples of mollusk diversity: (a) A polyplacophoran, *Chiton granosus*. (b) A pulmonate gastropod, *Monache syriaca*. Note that the mantle cavity can be seen through the thin shell. (c) A bivalve, the king scallop *Pecten maximus*. (d) A cephalopod, the common octopus *Octopus vulgaris*.

Source: a: Shutterstock, https://www.shutterstock.com/image-photo/edible-chiton-granosus-bay-tortugas-beach-1787529884; b: Leah Khanasvili, Original (used with permission); c: Shutterstock, https://www.shutterstock.com/image-photo/great-king-scallop-on-seabed-1962845185; d: Shutterstock, https://www.shutterstock.com/image-photo/octopus-vulgaris-cuvier-1797-cephalopod-octopodidae-1632380029.

live burrowed in the substrate. Many of them lack some of the classic molluskan characters, such as a muscular foot, and lack eyes and statocysts. Solenogastres mostly feed on hydrozoan polyps, whereas caudofoveates are detritivores or predators of large single-celled eukaryotes. Several hundred species are known, but this is probably an underestimate, since they are an understudied group.

Polyplacophora ("bearing many shells") is made up of the chitons, benthic creeping animals with eight longitudinally arranged shell plates. Aside from the repeated shell plates, their body-plan is fairly conservative with a mantle, visceral mass, muscular foot, and radula. Another unusual polyplacophoran feature is the existence in some species of numerous eyes, embedded in the dorsal plates. These eyes take the form of simple **ocelli**, sometimes with a crystalline lens. Close to 1000 species are known, as well as several hundred fossil species. Many live in tidal zones, attaching to the rocks in shallow water, but there are also deep-sea species.

Monoplacophora is a rare group. Until the middle of the 20th century they were known only as fossils, dating back to the Cambrian. Since then, some 30 species have been described, almost all found exclusively in the deep sea. They have a single domed shell covering the dorsal part of the body, hence their scientific name—literally, bearing one shell. Monoplacophorans have no eyes and bear simple sensory organs. They have eight pairs of dorso-ventral muscles that reach the foot, and repeated pairs of ctenidia and excretory organs.

Scaphopoda ("boat foot") the tusk shells. These are animals with a long and tapering almost crescent-shaped single shell that is open at both ends. Scaphopods normally live buried in the sediment, drawing water in through the anterior. Like many sedentary animals, they have lost some of the typical molluscan structures, and lack eyes or ctenidia. Scaphopods have a large mantle cavity, through which water flows. They use specialized tentacles known as **captacula**—often numbering in the hundreds—to collect small prey items from the sediment. The prey, usually large single-celled organisms, is brought to the mouth and crushed by a large radula. Some 500 species of scaphopods are known.

Bivalvia includes the well-known clams, mussels, oysters, and other similar animals. Bivalves are characterized by a shell with two articulated valves, hence the name. The two valves can close to isolate the animal completely from its environment, thereby providing protection from predators and other dangers. Adductor muscles keep the valves tightly closed, while ligaments open the valves when the adductor muscles relax. The valves open to allow feeding or release of gametes. Bivalves have an expanded mantle, which secretes the two-part shell. They normally feed by filtering food particles out of the water, with water entering through an inhalant opening and exiting through an exhalant opening. In some cases, these openings are extended as soft tubes, in which case they are known as **siphons**. Since bivalves feed by filtering small food particles that do not require processing, they have lost the radula. The ctenidia adopt a secondary role and function both as a breathing organ and as a filtration organ to trap food.

Some bivalves live buried in the sediment, with the foot functioning mostly as a digging organ. Some groups of bivalves have adapted to living on hard surfaces, and in these cases the foot functions as an adhesive organ, often by secreting a type of fiber, known as **byssus**, that aids in keeping the animal in place. There are close to 10,000 species of bivalves, found in almost all benthic habitats, from sandy shores to rocky tidal regions to the deep sea.

Cephalopoda is a small class in terms of species number—only about 700 extant species are known—but there are many thousands of fossil species. Cephalopoda includes some of the most complex and impressive members of Mollusca: octopuses, squid, and their relatives. Cephalopods have undergone a radical change in their body plan relative to other mollusks, as part of their transition to an active and predatory lifestyle. Their original dorsal–ventral axis has elongated significantly and has been functionally transformed into a novel anterior–posterior axis. The foot has expanded and split into a number of flexible muscular arms, sometimes with additional tentacles. The visceral mass also expands and the head region is highly elaborated. Most extant cephalopods do not have an external shell. Octopuses have lost the shell completely, whereas squid and cuttlefish have internalized it to function as an internal support element, that also has a role in maintaining bouyancy. However, fossil members of Cephalopoda, the ammonites and nautiloids, had well-developed coiled shells. The ancestral shelled state is still found in the few surviving members of the nautiloids. Cephalopods have a chitinous two-part beak, with food being processes both by the beak and by the radula.

Gastropoda is the largest molluskan class, with over 100,000 described species of snails, slugs, and their relatives. Gastropods have a generalized molluscan body plan with a few notable modifications. Their most obvious feature is the muscular ventral foot, which gives the class its name, which translates as "stomach foot." The character that distinguishes extant gastropods from all other mollusks is the existence of visceral mass **torsion** (Figure 15.3). The dorsal portion of the gastropod body and the ventral part rotate relative to one another. This rotation is ancestrally through 180°, bringing the mantle cavity opening and the anus over the anterior part of the body. In many groups of gastropods there has been a secondary detorsion of 90°,

bringing the opening to the side. Gastropods are sometimes divided based on the degree of torsion but note that this is a functional and not a taxonomic division, as these are paraphyletic groups. Prosobranch gastropods are those that maintain the ancestral condition of the mantle cavity and anus pointing anteriorly (prosobranch translates as "gills forward"). Opisthobranchs ("gills backwards") are those where the openings have rotated an additional 90° to bring the opening to the right side. Pulmonates are the land snails, where the mantle cavity has expanded to form functional lungs.

In addition, but independently of visceral mass torsion, many gastropods also have a coiled shell. The shell grows radially as the animal grows, providing an internal space into which it can retreat for protection. Many gastropods lack a shell, and it has been lost convergently several times, both in marine species and in terrestrial species.

Evolutionary history

Because of their mineralized shells, mollusks have an excellent fossil record. Fossils of all shelled molluskan taxa are known from nearly all relevant fossil strata. The earliest fossil that is identified as a putative mollusk is the late Ediacaran (555 million years ago) *Kimberella*. Putative aculiferans and monoplacophorans are the earliest definite mollusks in the fossil record and are found from the early Cambrian. Cephalopods first appear as fossils in the Late Cambrian, about 500 million years ago, and rapidly become the top marine predators. The nautiloids were dominant members of the nektonic communities throughout the Paleozoic and were replaced by ammonites in the Mesozoic. Ammonites went extinct in the end-Cretaceous mass extinction (which also wiped out the dinosaurs), whereas nautiloids remain with very low diversity to this day. Snails and scaphopods are common fossils in almost all eras, although scaphopod diversity today is much lower than it has been in the past. Bivalves rose to prominence as major components of benthic communities during the Mesozoic. Gastropods are the only molluskan group to invade the land. This occurred several times within this group, probably around the Permian.

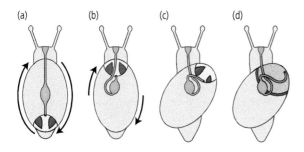

(a)   (b)   (c)   (d)

Figure 15.3 Gastropod torsion: The different evolutionary stages in gastropod torsion. (a) The hypothetical ancestral condition, with a digestive system (pink) stretching from an anterior mouth to a posterior anus. The gills (red) face backward. This condition is not found in any extant species. (b) The prosobranch condition. Following 180° rotation, the anus and gills move to the anterior. (c) The opisthobranch condition. Following an additional 90° rotation, the gills and the anus point to the right. (d) The pulmonate condition. The gill cavity has expanded and developed increased vascularization to function as a lung.

Source: Source: a: Photo supplied from Shutterstock: Stefan Ziemendorff (https://www.shutterstock.com/image-photo/edible-chiton-granosus-bay-tortugas-beach-1787529884); b: Photo supplied courtesy of Leah Khananshvili: Original (used with permission); c: Photo supplied from Shutterstock: Becky Gill (https://www.shutterstock.com/image-photo/great-king-scallop-on-seabed-1962845185); d: Photo supplied from Shutterstock: ennar0 (https://www.shutterstock.com/image-photo/octopus-vulgaris-cuvier-1797-cephalopod-octopodidae-1632380029).

CHAPTER 16

Skeletons and coeloms

Mechanical principles of motion

Before discussing skeletons, we have to go back to some basic concepts in physics. A **lever** is a simple machine that consists of a rigid structure or beam, which rests on a point known as the **fulcrum**. Force applied at a point along the beam, causes it to pivot around the fulcrum. If the force is applied far from the fulcrum on one side of the beam, it generates a stronger force close to the fulcrum on the other side. Think of lifting a rock using a shovel while pushing on the far end of the shovel. In contrast, if force is applied closer to the fulcrum, it needs to be stronger to generate the equivalent amount of movement. Think of lifting the same rock but pushing close to the blade of the shovel. A **joint** is where two beams meet at a shared fulcrum, allowing them to move one relative to the other. If one of the beams is fixed in place, force only has to be applied to the other beam to allow movement around the joint.

Returning to biology, the rigid beams can be elements of a hard skeleton connected by joints. The force is applied by muscles connecting to the skeletal elements, and the contraction of the muscles leads to movement of one of skeletal element relative to another. Muscles attached to elements of the skeleton close to a joint have to be stronger and can produce faster and more forceful movement, whereas muscles attached more distally do not have to be as strong and produce slower and weaker movement. Muscles need to be anchored at two sides to generate movement. In almost all cases, a muscle is connected to the two skeletal elements surrounding the joint. One element provides resistance and remains mostly stationary, whereas the other is moved by the muscle. Since joints need to both extend (open) and flex (close), joints normally

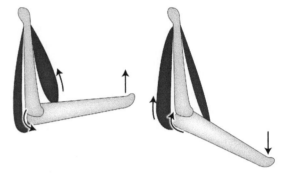

Figure 16.1 The action of muscles on a joint: Two bones (grey) connect at a joint. When the inner muscle contracts, the joint flexes (left). When the outer muscle contracts, the joint extends (right).

have two sets of muscles, one connecting to the skeletal elements on the inside of the joint and one on the outside. The internal muscle contracts, while the external muscle elongates in order to flex the joint while the external muscle contracts and the internal muscle elongates to extend the joint (Figure 16.1).

Internal and external skeletons

Skeletons are structural elements that provide support, muscle anchoring points, and protection. Skeletons are normally rigid, but as we will see, they can also be flexible, while still providing support and limited protection. Skeletons can be of two main kinds, **internal skeletons** or **external skeletons**. An internal skeleton (or **endoskeleton**) consists of individual elements, often in the form of **bones**, embedded within the body and providing internal support to the animal. Internal skeletons can also provide protection if they enclose a vulnerable organ (think

Organismic Animal Biology. Ariel D. Chipman, Oxford University Press. © Ariel D. Chipman (2024). DOI: 10.1093/oso/9780192893581.003.0016

of the vertebrate skull—Chapter 29). Echinoderms (Chapter 26) have an internal skeleton that lies close to the integument and thus also serves as a protective structure.

External skeletons (or **exoskeletons**) are usually composed of cuticle units or plates that surround the body and provide both physical protection and mechanical support. The exoskeleton can be a single unit, in which case it is usually referred to as a shell (as in snails). If it is made of multiple plates, the joints between elements of external skeletons usually include flexible or extendable membranes that allow the plates to move relative to each other without leaving a gap. Joints between elements of internal skeletons usually include connective tissue that allows the bones to move around the joint without abrasion. Muscles moving a joint in an internal skeleton attach to the external surface of the bone, whereas muscles moving a joint in an internal skeleton attach to the inner surface of the plate.

As a general rule, skeletons that are rigid require joints between discrete skeletal elements and muscles that attach to these elements on both side of a joint in order to enable movement. Skeletons that are less rigid can be composed of continuous units that can flex in response to the contraction of muscles located along the skeleton.

Several different materials make up skeletons in the animal kingdom. In many cases, calcium-containing minerals are the main component. Internal skeletons of vertebrates are composed of **bone** (note that the word "bone" is used both for the material composing the skeletal elements and for the elements themselves). The main component of bone is calcium phosphate or apatite. External skeletons of arthropods are composed of a complex polysaccharide known as **chitin**, often strengthened with calcium salts. The shell of some mollusks is a type of external skeleton (although usually with very limited movement) that includes calcium carbonate as the main mineral component. **Cartilage** is a variable type of connective tissue that is more flexible than bone or chitin, and acts as an internal skeleton in several taxa including sharks and squid. Cartilage also often includes calcium salts as part of the matrix. Finally, the cuticle of many phyla, including annelids and nematodes, is a more or less flexible external skeleton that is composed of various fibrous proteins arranged in layers to give protection and structural support.

Types of locomotion

Recall that in Chapter 8 we differentiated between **movement** as a general term for a change in position of any part of an organism, and **locomotion** as movement that leads to a change in location of an organism. Muscles (or sometimes contractile epithelia) are broadly responsible for both. Muscles don't necessarily need to activate joints to generate movement. In animals with flexible skeletons or with no significant skeleton, **longitudinal muscles** can attach to the body wall, and lead to movement of sections of the body wall relative to each other. **Circular** (or circumferential) **muscles** can contract in unison to generate a constriction of the body at a given point. Coordinated muscle flexion and elongation can lead to movements that provide locomotion. Oblique or diagonal muscles contribute to more complex and nuanced movement of the body.

We will give two examples of common types of locomotion that can be generated thorough the coordinated activity of body-wall musculature (Figure 16.2). In **peristaltic locomotion**, longitudinal and circular muscles contract together at a given point along the animal's anterior-posterior axis and relax at points anterior and posterior to

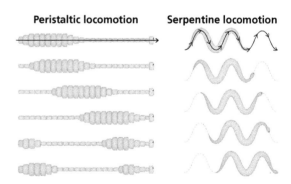

Figure 16.2 Different types of locomotion using body-wall musculature: Peristaltic locomotion (left) illustrated with an earthworm. The relative dimensions of the contracted and relaxed segments are exaggerated for emphasis. Sinusoidal locomotion (right) illustrated with a snake. The arrow indicates the direction of movement.

this. A travelling wave of contractions and relaxation allows the animal to move without lateral flexing of the body. This is the type of locomotion seen in earthworms. It is an efficient mode of locomotion when burrowing into the substrate. Peristaltic movements involving synchronized contractions of circular and longitudinal muscles are also the mechanism for moving food along the digestive tract in many animals. In **sinusoidal** or **serpentine** locomotion, longitudinal muscles on one side of the animal contract, while the muscles at the other side relax and elongate. Coordinated muscle contraction and relaxation causes the animal to move in a way that looks like a traveling wave. This is the type of locomotion used by snakes and many worms. It is an efficient mode of locomotion for moving along the substrate, and when aided by fins or paddles can also allow swimming in the water column.

Animals that have more rigid skeletons often move using external appendages: limbs or fins. Limbs are rigid structures that extend from the body wall and are often composed of several rigid elements connected by joints. Coordinated movement of the limbs against the underlying substrate pushes the animal forward, in a type of locomotion usually known as walking (see Chapter 8 for additional terms). Fins (or paddles) are flattened structures extending laterally from the body wall that push against the water, to allow the animal to swim.

Coeloms

A **coelom** is a type of internal skeleton that is composed of a body cavity filled with fluid under pressure. It is often referred to as a **hydrostatic skeleton**, meaning a skeleton that utilizes the incompressibility of water. Water is not compressible, but it is fluid, allowing a water filled cavity to be flexible, while still providing resistance. In many soft-bodied animals that lack a rigid skeleton, the coelom is the main internal support structure that allows muscles to generate movement.

Coeloms are derived from cavities within the mesodermal layer. A complete coelom is a cavity that is bounded by mesoderm on all sides (Figure 16.3A). It can be continuous along the main body axis, or be composed of individual repeating cavities. This is the type of coelom found in most annelids, in chordates and in some mollusks. A type of cavity known as a **pseudocoelom** (Figure 16.3B) is found in representatives of some phyla including nematodes and rotifers. It is distinguished from a complete coelom by the fact that it is bounded by mesoderm externally and surrounds the endodermal digestive tract directly, with no mesoderm in between. Some phyla, such as the platyhelminths we met in Chapter 13 lack a coelom altogether and are known as **acoelomate** (Figure 16.3C).

In the older zoological literature, a great deal of attention was given to the type of coelom present in different animals, and this was used as an important tool for classifying phyla. The common idea was that acoelomates are the more primitive phyla, that pseudecoelomates evolved from them, and that animals with a true coelom represent the most advanced phyla. Current phylogenies (see Box 16.1) show this idea to be wrong. Coeloms are believed by some researchers to be ancestral to Bilateria and to have been lost in several phyla and reduced in others. The main reason for reduction or loss of coeloms appears to be the evolution of smaller size (see Chapter 19 for more consequences of smaller size.)

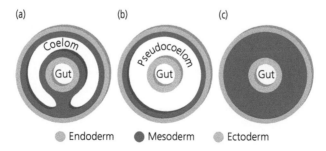

Figure 16.3 The organization of types of body cavities: Schematic cross section through the body of an animal with a complete coelom (a), a pseudocoelom (b), and no coelom (c).

Box 16.1 Old phylogenies, new phylogenies

Attempts at classification of animal diversity are as old as the study of animals themselves. As far back as the fourth century BC, the Greek philosopher Aristotle developed a preliminary system of animal classification. The eighteenth century Swedish naturalist Carl Linnaeus is credited as the inventor of modern classification and is the father of the system named in his honor, which we met in Chapter 1. By the early nineteenth century, different classification schemes were suggested by several naturalists, usually using a hierarchical and often dichotomous system, presaging the modern notion of bifurcating phylogenetic trees (see Chapter 4). The first graphical depiction of a phylogenetic tree is Charles Darwin's famous "I think. . ." diagram from his notebooks. The only illustration in Darwin's *On the Origin of the Species* is what we would today call a phylogenetic tree.

Early phylogenies were largely intuitive and were not based on an underlying theory or series of rules. They were often created by lumping together taxa based on perceived similarities, and in many cases included implicit assumptions about the order of appearance of structures throughout evolutionary history. These phylogenies tended to assume a gradual evolution from simple to complex structures, and thus clustered "simple" organisms at the base of the tree and "advanced" organisms near the top.

Throughout the late twentieth century, there appeared several competing theoretical frameworks for creating phylogenies. When these were combined with molecular data, a very different picture of animal phylogeny began to emerge. The new phylogenies rejected the notion of progression from simple to complex organisms in moving up the tree. They broke up several accepted clusters of "primitive" animals such as the infamous "aschelminthes" and the classification of animals based on presence or absence of a coelom. Conversely, the new phylogenies of the late twentieth century brought together groups of organisms that were not traditionally thought to be related, most notably erecting Spiralia/Lophotrochozoa (see Chapter 13) and Ecdysozoa (see Chapter 20). While it is important to remember that every phylogenetic tree is a hypothesis, the new phylogenies are supported by an ever-increasing amount of data from various sources. With time, there are fewer unresolved sections of the tree, as more data is accumulated and as the phylogenetic reconstruction algorithms improve.

CHAPTER 17

Annelida

An introduction to the annelids

Annelida is a large phylum of worms, with over 20,000 species, distributed worldwide. The name comes from the Latin *annulus* or "ring," referring to the externally visible repeated rings that make up the annelid body. Most annelids are found in the sea, either as free-swimming or crawling species, or as various sedentary, burrowing, or tube-dwelling worms. Some annelids are found in fresh water or in damp soil, such as the familiar earthworms. One group of annelids is primarily composed of blood-sucking ectoparasites—the leeches. Annelids have a generally more conservative body plan than some of the phyla we have met up to now, but a few lineages have diverged from it significantly. Nonetheless, annelids have adapted to numerous lifestyles and feeding modes and are common members of nearly all marine ecosystems as well as terrestrial soil ecosystems. The annelids are important members of the group of spiralian phyla. Some other members of this group, which are not discussed, elsewhere are described in Box 17.1.

Annelid body organization

Annelids are bilaterally symmetrical triploblastic worms, with a true coelom and a segmented body. **Segmentation** is a type of body organization that we will discuss in detail in Chapter 18. For now, we will define it simply as a body plan that is composed of repeating units, each containing elements from a number of organ systems. Most annelids are visibly segmented externally. The internal anatomy is also segmented, with the ancestral condition including segmentally repeated coelomic pouches, each matched with segmental muscles, segmental ganglia, and segmental excretory organs (Figure 17.1). Some annelids have secondarily lost the segmental body organization, either externally or internally (e.g., echiurans).

Many annelids have lateral external bristles or **chaetae** (**setae** in some texts). These bristles were traditionally used to divide annelids into polychaetes ("many bristles") and oligochaetes ("few bristles"), with the leeches considered to be a third group. As we will see later in this chapter, this division does not stand up to phylogenetic scrutiny, but is a useful distinction. Polychaetes normally grow their bristles from lateral extensions of the body wall known as **parapodia**, which are used for locomotion and often function as gas exchange organs (Figure 17.2).

The anterior part of the annelid body is known as the **prostomium** and is a non-segmental region that carries many of the sense organs. This is followed by the **peristomium** which holds the mouth, accessory feeding structures, and additional sense organs. These two units can together be considered to be the annelid head. The level of development and specialization of the head region is variable and can be linked to the level of motility and lifestyle, with active predators having well-developed heads, and sedentary suspension feeders having minimal heads. The bulk of the body is made up of serially repeated segments, often similar to one another with minimal specializations. Tube dwelling and sedentary forms often have segments that are regionally specialized, a condition known as **heteronomous segmentation**. The posterior-most non-segmental region is the **pygidium**.

Annelids can reach significant sizes for worms, some species being as long as several tens of centimeters, and in extreme cases even longer than a

Organismic Animal Biology. Ariel D. Chipman, Oxford University Press. © Ariel D. Chipman (2024). DOI: 10.1093/oso/9780192893581.003.0017

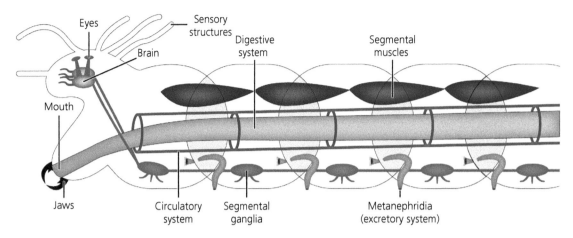

Figure 17.1 Main elements of the annelid body plan: Lateral view of a generalized annelid, combining elements from different annelid groups.

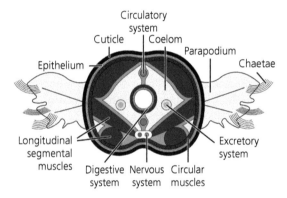

Figure 17.2 Transverse section through a generalized annelid: General arrangement based on polychaete annelids.

meter. The increase in size can be attributed to an efficient circulatory system coupled with a robust external cuticle, muscular body wall, and a coelom that together provide support.

Annelid life history and reproduction

As a rule, annelids are sexually reproducing, but there are numerous variations on this that can be found within the group. Most marine polychaetes have separate sexes and gametes are released into the surrounding water where they are fertilized. In some polychaetes, reproductive individuals undergo a morphological transformation to become **epitokes**. The epitokes swell with gametes and modify their locomotory and sensory organs,

to allow them to swim up into the water column, where they release their gametes. In some taxa, such as the nereids, the entire worm transforms into an epitoke, whereas in other cases, such as syllids, it sheds groups of gamete-filled segments that are capable of autonomous locomotion. There are even cases where these shed segments grow their own heads and tails, to essentially become asexually produced sexually reproductive units. Sexual maturation into epitokes is finely synchronized within a population, so that many male and female epitokes swarm and release their gametes at the same time, increasing the chance of fertilization and placing the embryos high in the water column where they can begin their planktonic larval life. The larvae are trochophore larvae as typical for many phyla within Spiralia.

Internal fertilization is found within the leeches and oligochaetes, collectively known as Clitellata, for their unique reproductive structure, the **clitellum**. Clitellates are hermaphrodites, and normally mate reciprocally, with two individuals lining up facing opposite directions, so that the male sperm opening or **gonopore** lines up with the female sperm collection vessel, the **spermatheca**. Posterior to the male and female reproductive organs is a thickened region called the clitellum that is characterized by a glandular epithelium. This epithelium secretes a **cocoon,** formed by a desiccation-resistant outer cover and a fluid interior. The cocoon moves forward across the reproductive organs. As it moves, it passes over the female gonopore, picking up ova, and then over the spermatheca, picking up sperm. Thus, fertilization occurs within the cocoon, which eventually moves through the head and is shed, becoming a sealed egg case. The cocoon is deposited under the soil or in some other protective environment. The eggs develop in the cocoon and hatch from it. Clitellates have direct development, and the hatchlings resemble miniature adults.

Many annelid taxa exhibit asexual reproduction as well. There is a diverse array of modes through which annelids reproduce asexually. One mode is through transverse fission, where a single worm divides and becomes two. Each half then regenerates to become a full worm. The regeneration process may begin before the two resulting worms are fully split. Another mode is through segment shedding, where individual segments are released from the posterior, each developing into a complete worm. There are even extreme cases where annelids reproduce by multiple fragmentation, where the worm breaks up into many pieces, each developing into a full worm. The ability to develop an entire animal out of cut pieces is linked to many annelids' extensive regenerative abilities. Indeed, most annelid taxa are capable of regenerating severed parapodia, missing segments, and even missing heads. It has been argued that asexual reproduction could not have evolved without a pre-existing mechanism for whole-body regeneration, although there is still some debate about this question. If this argument is true, once regeneration was used for asexual reproduction there was a strong selective pressure to evolve ever more efficient modes of enhanced regeneration and concomitant asexual reproduction.

Feeding and digestive system

Annelid digestive systems are normally straight and simple, with specific modifications based on feeding habits. The digestive system is typically divided into three regions: the foregut, midgut, and hindgut. The foregut includes the mouth and pharynx and is lined with cuticle. The cuticle produces accessory feeding structures such as sclerotized jaws, which are often hard and mineralized.

In some annelids, the transition between the foregut and midgut develops into a muscular crop where food is stored and processed.

The midgut is lined with endodermal epithelium. This region can be roughly uniform or functionally divided into stomach and intestine, with digestive and absorptive functions respectively. In earthworms, there is a dorsal expansion, known as the **typhlosole**, that increases the internal surface area of the intestine. In other cases, surface area is increased by less prominent grooves and wrinkles. The hindgut is relatively short and includes only the rectum, which opens at the anus.

In a taxon as diverse as Annelida, it is no surprise that we can find nearly all types of feeding mode. Actively hunting predators as well as ambush predators that tackle prey larger than themselves are common, especially among marine groups. Many sedentary species feed from suspended sediment, while ground crawling and burrowing species are detritivores. Herbivory is rare but exists. Some species show various combinations of all of the above. Last but not least, many leeches are ectoparasitic blood feeders.

Predatory annelids are common in all marine environments. Some are active predators that swim or crawl along the surface hunting for prey. Others are ambush predators, lying buried in the sand emerging to grab prey and pull it into their burrow. The pharynx of predatory annelids is often armed with sharp strong teeth that can hold on to prey. The teeth are usually chitinous, and in some cases are strengthened with iron and other minerals. Many predatory annelids have an **eversible pharynx** which allows them to increase their gape enough to trap large items of food. In glycerids (bloodworms), there is an eversible proboscis adorned with jaws connected to venom glands, used to envenomate and paralyze their prey.

Most sedentary annelids are suspension feeders. They live in tubes or burrows and either draw water into their burrow, filtering out food particles from the water passing by their mouths, or use a diversity of tentacles, arms, nets, or fans to catch particles from the surrounding water and bring them to their mouths. Some tube- and burrow-dwelling annelids have evolved specialized structures that remove digestive waste and prevent their homes from becoming fouled. In other cases, the burrow is U-shaped, so that the anus is close to the surface rather than emptying into the burrow.

Deposit feeding and detritivory are also common in marine and terrestrial annelids. Some species crawl along the sea floor, eating any organic matter they find along the way. Other deposit feeders are more specific and use chemical cues to find suitable food. Earthworms ingest the soil they pass through as they move under the surface and extract organic matter from it, passing the undigestible elements of the soil through the gut to the anus.

The leeches have become specialized blood feeders (although some species are secondarily predators or scavengers). They have one or two adhesion glands or "suckers" that allow them to attach to their host, where they remain until they are fully engorged with blood. This mode of feeding requires significant modification of their body plan in general and of their digestive system in particular. Leeches have lost the septa separating segments, to allow them to inflate their bodies. They also have a highly developed excretory system that helps them get rid of the excess liquid from the blood rapidly and efficiently. Another specialization for blood-feeding in leeches is the secretion of antico-agulants into the wound they create on the host, to allow blood to flow freely, and analgesic and anti-inflammatory compounds to prevent the host from noticing their presence.

Some annelids, including siboglinids and others have lost the digestive system entirely in the adult stage. These animals live in anaerobic or sulfur-rich habitats and draw all of their nutrition from symbiotic bacteria.

Circulatory system

The most common type of circulatory system in annelids is a closed circulatory system. There is normally a dorsal vessel that carries blood anteriorly and a ventral vessel that carries blood posteriorly. Segmental minor vessels carry blood from the main dorsal vessel to a network of minor vessels that supply all tissues with oxygen and digestive products and collect carbon dioxide and waste back to the ventral vessel. Pumping of the blood is both through muscles in the vessel walls and contraction of body wall muscles.

Oxygen and carbon dioxide are interchanged directly through the body wall and the gut in small annelids and in most clitellates. In marine species bearing parapodia, these structures also function as gills. In larger and more active species, the external surface of the parapodia is enlarged to allow enough exchange surface area. The parapodia are supplied with numerous blood vessels that carry oxygen back to the main vessels. Freshwater annelids normally lack parapodia. Those inhabiting oxygen-poor habitats have therefore often evolved extensions from the body wall, usually laterally or at the posterior end to improve gas exchange.

While the layout of the circulatory system described above holds for most annelid taxa, there are many variations. There are some large and active species that do not have well developed circulatory systems, and some of the transfer of oxygen is done via the coelomic fluids. Many sedentary annelids have evolved enlarged sinuses that pump blood actively, presumably as a replacement for blood pumping by the body wall muscles found in more mobile species.

The blood of some annelids contains blood cells with hemoglobin, chlorochruorin, or related pigments. In other cases, respiratory pigments are dissolved in the blood fluid without being packaged in cells.

Muscles and movement

The segmental body plan of annelids allows for efficient locomotion. Longitudinal muscles connect adjacent segmental external rings, allowing coordinated movement. A combination of these longitudinal muscles and circular body-wall muscles allows for serpentine locomotion or peristaltic locomotion (see Chapter 16), depending on the life style. Some annelids are capable of swimming, fast-crawling, or slow-crawling, with the difference between the locomotory modes depending on different rates of muscle contractions and a modulation of the coordination among the muscles.

All types of locomotion can be found within annelids, from active swimmers to crawlers and burrowers. Many taxa are sedentary, but none can be considered to be truly sessile. Tube-dwelling annelids move within their protective tubes, but do not normally leave them or replace them. Most marine annelids, regardless of their mode of locomotion as adults, have planktonic larvae.

The arrangement of the segmented body is closely linked with the mode of locomotion. Most active annelids have **homonomous segmentation**, meaning all segments are similar in size and shape. Many tube dwellers and burrowers have heteronomous segmentation, with different regions being specialized for different functions—digging, holding the animal in place, feeding, and others. Segmentation has been completely lost in some sedentary annelid taxa. The lack of segmentation led to these taxa being originally identified as separate phyla, notably the echiurans and sipunculids.

Nervous system

The central nervous system of annelids is composed of a series of ventral segmental ganglia. The ganglia are connected by paired cords, resulting in a typical ladder-like nervous system. At the anterior of the nervous system is a circumesophageal **commissure** (similar to what we saw in Mollusca in Chapter 15). The cerebral ganglion, which forms the annelid brain, connects to the ventral ganglia via the circumesophageal commissure. The brain is ancestrally housed in the prostomium, although it has been shifted backward in several lineages. Active species have an enlarged brain that can be divided into three regions, the forebrain, midbrain, and hindbrain, each innervating different structures in and around the head.

Many annelids have eyes, but almost none have well-developed camera-eyes or lens-eyes. The prostomium often houses one or more pairs of anterior ocelli, which develop from pigment patches in the trochophore larva. Pigmented eyes are sometimes found along the body and not just in the head. Conversely, burrowing and tube-dwelling species often lack eyes entirely.

In contrast with their rather simple visual structures, annelids tend to have highly complex tactile and chemical senses. Chaetae act as tactile sensory structures, together with various **palps** or antennae and surface tactile receptors. Chemical receptors are found in almost all parts of the body in different

annelid species, including on the palps and parapodia. Statocysts are found in many annelids, but there is usually only a pair or a few pairs of these structures near the head.

Annelid diversity and taxonomy

Recent molecular work has led to a significant restructuring of the annelid tree. The relationships among the different taxa are in flux, and the placement of family and ordinal taxa into higher taxonomic groups is often controversial. The main branch of Annelida is now divided into two clades Errantia and Sedentaria, with a number of smaller lineages lying outside this major division. Both clade names are historical taxonomic names that were resurrected recently in light of the revised view of annelid phylogeny. Errantia in its modern sense includes several lineages of motile ("errant") worms, whereas Sedentaria includes many lineages of sedentary worms, including some traditionally believed to be outside of Annelida altogether.

The traditional division into three classes, Polychaeta, Oligochaeta, and Hirudinae (leeches), has thus been abandoned. Polychaetes are now seen as a generalized, paraphyletic grade, composed of several distinct lineages. The leeches and oligochaetes (including earthworms and several freshwater taxa) are now united under Clitellata, characterized by hermaphroditic reproduction with a clitellum as described above, which is just one of many lineages within Sedentaria.

We will survey a representative sampling of taxa—some traditionally considered to be families and some orders—without committing to higher level taxonomy (see Figure 17.3).

Errantia includes the majority of annelids species. Most errant annelids are homonomous active worms. Nereidae (bristle worms) includes long worms adapted for swimming and fast crawling including the well-known developmental model species *Nereis* and *Platynereis*. Eunicida includes about 1000 species with well-developed jaw musculature, some of which are large ambush predators like the sand-striker. Phyllodocida include nearly 5000 species of mostly benthic crawling worms, including sea mice, scale worms and their relative.

Figure 17.3 Examples of annelid diversity: (a) An errant nereid, the sandworm *Perinereis*. (b) A sedentary sabellid, the fanworm *Sabella spallanzanii*. (c) A clitellate, the earthworm *Lumbricus terrestris*. (d) A clitellate, the medicinal leech *Hirudo medicinalis*.

Source: a: Photo supplied from Shutterstock: Somprasong Wittayanupakorn (https://www.shutterstock.com/image-photo/close-sandworms-perinereis-sp-polychaeta-isolated-1274989057); b: Photo supplied from Shutterstock: Andre Labetaa (https://www.shutterstock.com/image-photo/spiral-tubeworm-sabella-spallanzanii-mediterranean-sea-2212038135); c: Photo supplied from Shutterstock: D. Kucharski & K. Kucharska (https://www.shutterstock.com/image-photo/earthworm-lumbricus-terrestris-171009224); d: Photo supplied from Shutterstock: Mit Kapevski (https://www.shutterstock.com/image-photo/medicinal-leechhirudo-medicinalis-743110495).

Sedentaria includes a diversity of tube worms, burrowing worms, and other sedentary taxa. About 13,000 species are included in this group with nearly 20 families, of which we will mention but a few. Most members of this group have a body that is heteronomous at least to a certain degree, and they tend to have less highly developed sensory organs. A well-known family is Serpulidae, consisting of tubeworms that secrete calcareous tubes. Serpulid tubes can be found on many hard substrates in the sea, including on molluscan shells, and on manmade object such as piers and ship hulls. Arenicolidae is not a very species-rich family, but includes the familiar lugworms, who live in J-shaped burrows and have a body clearly divided into a number of functional regions. Echiuridae (spoon worms) was once considered a separate phylum. Echiurids have lost all traces of external or internal segmentation and have a long feeding appendage that may be several times longer than the worm itself.

Siboglinidae was also considered a separate phylum due to their lack of digestive system. Clitellata, united by several reproductive characters and molecular analyses, includes about 1000 species of leeches, arranged into several families, close to 4000 species of earthworms and related forms, and about 1000 species of freshwater worms, including the Naididae, some of which lack digestive systems as adults and some that are commonly used as fishing bait.

Evolutionary history

Surprisingly, given their soft body, annelid fossils are known from the Cambrian, when they already had the typical annelid body plan and had diversified into the main morphological classes we know today. There are long swimming forms, benthic crawlers, and even tube-dwelling species known from their early fossil record. Invasion of land by earthworms occurred late relative to other terrestrial lineages, probably only in the mid-Mesozoic, about 150–200 million years ago.

Box 17.1 More spiralian phyla

Before we leave Spiralia, we will mention a few more phyla that are worth being acquainted with. Most of these are grouped under a clade known as Lophophorata, named for the presence of a typical structure, the lophophore. This is a specialized feeding appendage, usually including a ciliated horseshoe or ring-shaped structure at the end of a long arm. The lophophore collects food particles from the water and brings them to the organism's mouth.

Bryozoa—The moss animals. There are over 6000 bryozoan species known, and more than double that known fossil species. Bryozoans are sessile colonial animals. The individual zooids are on the scale of 1mm in size.

Each zooid has a retractable feeding arm that delivers food to a U-shaped gut. The colonies can form diverse species-specific shapes ranging from branched coral-like structures to flat lace-like colonies that encrust hard surfaces. Bryozoans are found in both freshwater and marine environments.

Brachiopoda—Lamp shells. Often mistaken for bivalves, brachiopods are a distinct phylum, with a very different body plan to that of mollusks. The two valves that cover and protect the brachiopod's soft anatomy are dorsal and ventral valves, in contrast with the left and right valves of bivalves. Inside the protective shell are a long extendible lophophore and a relatively simple body with a minimal circulatory system, a simple U-shaped digestive system, and a reduced nervous system. Brachiopods have an extensive fossil record and were much more diverse in the past. There are only about 450 species known today, but there are some 12,000 known as fossils, found in almost all marine fossil deposits.

Phoronida—A small and poorly known phylum, with only 15 described species. It includes sessile marine lophophorates with no shell, but with a chitinous tube in which they live. Phoronids are worm-like, ranging in size from 1 to 45cm. Although the number of species is small, some species are distributed very widely.

Nemertea—Ribbon worms. Unlike the other phyla in this list, nemerteans are not lophophorates. They are active swimmers with an elongate flat body, reaching lengths of over one meter. Many nemerteans are colorful with striking patterns. Most are predators or detritivores, often swarming to consume large dead animals rapidly. Nemerteans display a mix of simple and complex characters, including a rudimentary coelom, but a closed circulatory system. They have long, differentiated digestive systems, multilayered integument and a centralized anterior nervous system, but a protonephridium-based excretory system. Despite their length, nemerteans are unsegmented. However, like the segmented annelids, asexual reproduction through fission is common. Nemerteans are characterized by an anterior proboscis, a complex extendable organ, which may or may not be connected to the digestive system. Its main function is prey capture.

CHAPTER 18

Segmental organization of the body

What is segmentation?

We have already come across segmented body plans in discussing the annelids. **Segmentation** is a morphological phenomenon in which the body is composed of repeated structures along the anterior posterior axis. The term "segmentation" is also used for the developmental process that forms the segmented body. Segmentation as a morphological phenomenon and as a developmental process has been studied and discussed extensively for several decades. There is therefore a rather extensive body of literature discussing the topic. In this chapter, we will discuss what it means to be segmented, how and why segmented body plans arose, and the diversity of these body plans in the animal kingdom.

Repeated units are fairly common in animal body plans. Units can be repeated along a main body axis or scattered throughout the body. Segmentation refers to the repetition of units or structures along the anterior–posterior axis. Some authors use the term **metamerism** or **metameric organization** to describe the repetition of units from a single organ system and reserve the term segmentation for a narrowly defined phenomenon at the level of the whole organism. In this view, segmentation is a type of body plan organization in the same way that symmetry is a type of organization. We will adopt this narrow definition and use the term segmentation to describe a body plan that has a coordinated repetition of structures from different organ systems (see Figure 18.1). These can include muscles, skeletal elements, ganglia, excretory structures, external appendages, and others. Each segment in this body plan is a complex unit that includes

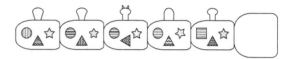

Figure 18.1 Schematic representation of a segmented body plan: In this illustration, each segment (repeated unit) includes elements from different body systems, each system illustrated as a single shape (triangle, circle, or star). The elements appear in each segment, but they may be modified (different orientation or internal pattern). Each segment also carries an appendage, which may have a slightly different shape in each segment. The posterior (right hand) segment indicates an undifferentiated embryonic segment.

elements from different systems. In segmented body plans, there may be sharp, externally visible borders between segments, but this is not always the case.

Segments can develop during embryonic development or they can be added post-embryonically as part of the organism's growth process. Much of the experimental work on segmentation has focused on embryonic segmentation, and we have a much better understanding of the processes that occur there. However, it is important to remember that many segments in many organisms do not form embryonically. During early embryonic development, there is a series of genetic interactions that define the borders between the nascent segments. The segments are thus first generated as distinct but undifferentiated units, with the elements of different systems initially indistinguishable. After the segments are formed, they differentiate to give different types of tissues and structures belonging to different organ systems. In post-embryonic segmentation, there is usually no early undifferentiated stage.

Organismic Animal Biology. Ariel D. Chipman, Oxford University Press. © Ariel D. Chipman (2024). DOI: 10.1093/oso/9780192893581.003.0018

Why be segmented?

Segmented body plans are highly adaptable, since individual segments can evolve independently to give segment-specific specialization. We can say that the segmented body plan is an example of **modular** organization and that each segment represents an independent morphological and evolutionary module. Modularity gives an advantage over time, because it allows adaptation of specific units (e.g., segment specialization) without affecting other units. However, this advantage to segmentation only comes into play after segmentation has evolved. It does not in itself explain the selective pressure that led segmentation to evolve in the first place.

There is no clear evolutionary explanation for why segmentation initially evolved. One hypothesis is that segmentation allowed more efficient movement. As we saw in annelids, the existence of repeated longitudinal muscles along the body axis allows for different types of locomotion and provides better coordination between segments as they bend relative to each other. However, this is not the case in arthropods, which have more rigid bodies and normally use a very different mode of locomotion. The selective advantage for the evolution of segmentation in arthropods may have initially been linked to the existence of segmentally repeated appendages, which allow locomotion by walking without bending the body.

A more general hypothesis for the origin of segmentation is that it may be a side effect of an elongation of the body. As the ancestors of segmented animals got longer, several systems were divided into repeated elements as they had to be spread out along the body. An organization in which the repeated elements are aligned and develop via a common developmental process is more efficient and may have been the driving force behind the evolution of whole-body segmentation.

The diversity of segmentation

In terms of species number, segmented animals are much more common than unsegmented ones, but this is because segmented taxa tend to be more species-rich—probably an outcome of their high adaptability and modularity. If we look at the level of phyla, we find that there are only three phyla with whole-body segmentation, and a few others that display varying levels of metamerism. The segmented phyla are Annelida (Chapter 17), Arthropoda (Chapter 21), and Chordata (Chapter 27). These three phyla belong to the three main branches of Bilateria: Spiralia, Ecdysozoa, and Deuterostomia, respectively.

Segmentation is manifested differently in each one of the segmented phyla. Annelid segments are sharply defined, with externally visible rings and corresponding internal septa separating segments. Arthropod segments are clearly visible externally, but are not separated internally, and the borders between them are often not clear. In chordates, segmentation is not visible externally and there are no internal borders between segments. Nonetheless, chordates have clearly repeated units of several systems that form in embryonic development and are maintained through to adulthood and should therefore be seen as segmented even though this is not as obvious as in the other two segmented phyla.

The segments of an organism may be similar to one another (as in earthworms or centipedes)—a phenomenon known as homonomous segmentation—but this is actually not very common. In most segmented organisms, the segments vary along the body, and in extreme cases, every single segment may be different. Each segment ancestrally includes elements of all the repeated organ systems, but some segments may lose some of these elements. Thus, in arthropods, we will find some segments with appendages and some segments without appendages. Several segments may be fused with some or all of their constituent elements appearing only once in the fused structure and other structures revealing the multi-segmental origin. For example, in millipedes, pairs of segments are fused into **diplosegments** that include only a single skeletal element but two pairs of walking limbs.

The segmental body plan can be lost completely or partially within lineages that originally displayed it. As we saw in Chapter 17, segmentation has been lost several times within annelids, usually as an adaptation to a sedentary mode of life. Several arthropod taxa display fusion of segments and loss of external segmental borders, for example spiders where external body segmentation is all but gone. One group of parasitic arthropods, the parasitic barnacles, have lost segmentation entirely in the adult stage. There are no known cases of loss of segmentation within chordates.

Segment number can be fixed or variable within a taxon or within a species. There are cases of higher-level taxa in which all (or almost all) members of the taxon have a conserved number of segments, for example, insects or leeches. As a general rule, animals with large segment numbers tend to have more variation within the species as is the case with many errant annelids and with some centipedes. Segments can be formed during embryonic development or as part of a post-embryonic process. The formation of all segments in the embryo is known as **epimorphic** segmentation, whereas the addition of segments post-embryonically is known as **anamorphic** segmentation. In anamorphic segmentation, the number of segments varies with age, with older individuals having more segments than young individuals. In some cases, segments continue to be added throughout the organism's entire life (e.g., in many millipedes). Marine annelids nearly all have anamorphic development, whereas terrestrial annelids tend to have epimorphic segmentation. Chordates all display epimorphic segmentation, with no exceptions. Arthropods have a mix of anamorphic and epimorphic development, sometimes even between closely related taxa.

The origins of segmentation

Given that segmentation is found in all three major branches of the animal kingdom, the obvious question is whether it arose once in the common ancestor of these three phyla or in each phylum independently. In other words, is segmentation homologous or convergent? If it is homologous, we must conclude that the common ancestor of all Bilateria, a hypothetical animal often referred to as **Urbilateria** (see Chapter 11), was segmented. In the early

2000s, there was a scientific debate over the nature of Urbilateria, with some scientists suggesting a scenario with a complex Urbilateria and others arguing for a simple Urbilateria (Figure 11.2). Under the complex scenario, Urbilateria is posited to have been a segmented animal. This scenario was based mostly on the existence of many conserved genes involved in segmental development in different phyla (see Chapter 32). Under the alternative simple scenario, Urbilateria was similar to a flatworm, with a simple gut, a minimal central nervous system and circulatory system, and an unsegmented body. This scenario is based on the assumption that the first bilaterians were small animals that did not have the need for many of the complex organ systems. From this simple shared ancestor, convergence in gross morphological features is very common, and there are a number of shared design principles that drive the evolution of animal body plans. While the debate has not been settled conclusively, additional data has shifted the view toward the simple scenario, and it is at least the opinion of the author of this book, that the actual Urbilateria was more similar to the one envisaged by the simple Urbilateria scenario. This means that segmentation is not homologous among the three phyla with whole body segmentation, and it must have evolved in each one of them independently.

Going back to the similarities between the development of the segmented body plans of different phyla; when we look at the details, the differences among the segmented phyla are greater than the similarities. Many of the shared genes are genes that are involved in numerous developmental processes, not only in segmentation. Many of the similarities in process are due to shared principles of segmental organization and are not indicative of shared ancestry. Looking at the fossil record of the three phyla, there is no evidence that they were segmented at the earliest stages of their evolutionary history, and at least in some cases, we can see the gradual appearance of whole-body segmentation throughout the fossil record of these phyla.

CHAPTER 19

Size and complexity

Being the right size

A common theme in science fiction is a person or other animal that changes size dramatically, either growing or shrinking, but still has the same proportions. This type of **isometric** change in size would not work in the real world. Physics and mechanics work differently at different scales, so that organisms of different sizes must have different proportions in order to continue functioning. Growth over evolutionary time scales, as well as changes in size throughout an organism's growth and development tend to be **allometric**. The main reason for allometric growth is the laws of geometry. Let's first demonstrate with a simple example of a cube with an edge length of 1 cm. The volume of the cube is 1 cubic cm (cm^3) and the area of its sides (faces) is 6 square cm (cm^2). If we increase the length of each side to 2 cm, the volume becomes 8 cm^3, while the area of its faces becomes 24 cm^2. With a 2-fold increase of the linear dimensions of the cube, its surface area grows 4-fold and its volume grows 8-fold. Thus, the ratio between the surface area of the cube and the volume of the cube grows more slowly than the linear growth. This fundamental property of three-dimensional objects has profound implications for organismic size.

Shifting back to biology; as an animal grows (either over evolutionary time or over its lifetime), its volume and mass increase more quickly than its surface area. The number of internal cells increases more rapidly than the number of cells facing the environment. This influences many physiological properties. Let's give a few examples: If an animal absorbs oxygen through its body surface, then as it grows, the amount of oxygen each cell in the body receives decreases, as there are more internal cells

for each cell on the surface. Physiological processes in the cells generate excess heat. This heat is usually dissipated through the body surface. As an animal grows, less heat can be lost via the surface. Conversely, if an animal needs to maintain a higher temperature than its surroundings, a smaller animal would have to invest more energy into generating heat than a large one, since heat loss is greater at smaller volumes.

Size also affects many mechanical properties. The amount of load a skeletal element can carry is proportional to its cross-sectional area. As animals increase in size, their mass increases more rapidly than linear growth, so that skeletal elements have to be thicker in cross section relative to the size of the animal. Similarly, muscles that move the skeleton have to be more robust as size increases. Larger animals are relatively more muscular and have relatively thicker skeletons—internal or external—than smaller animals (ignoring extreme cases of very long and thin worms, where growing in size doesn't necessarily affect all dimensions). The physics of locomotion are also size-dependent. When moving in air, smaller animals have less momentum (a product of mass and velocity) than larger animals and are therefore more agile, being able to change direction with less effort. Small animals can jump to a height many times their body length (e.g., grasshoppers), whereas very large animals usually can't jump at all (e.g., elephants). Using a similar line of argument, we can understand why large flying animals have wings that are larger relative to their body size than their smaller flying relatives.

Moving in water is affected by size because of the physical properties of fluids. Fluid mechanics is more complex and less intuitive than the simple geometry we described above, but it has significant

Organismic Animal Biology. Ariel D. Chipman, Oxford University Press. © Ariel D. Chipman (2024). DOI: 10.1093/oso/9780192893581.003.0019

consequences for aquatic locomotion. At smaller sizes, water appears more viscous. Very small animals (at scales of a centimeter or less) have to invest significantly more effort into moving forward and cannot rely on inertia to keep them moving. Large aquatic animals tend to evolve streamlined hydrodynamic shapes, whereas in small animals the shape is less important. Many marine larvae have non-streamlined shapes, with projecting appendages and extensions that would generate significant drag if they were scaled up, but their effect is negligible at small scales because of the general viscosity. Fluid mechanics are important in aerial locomotion but are significant mostly at very small sizes. Small insects, such as bees, fly by relying on generating turbulence with their wings, whereas large insects, such as dragonflies, and most birds rely on lift generated by the curved surface of their wings.

Physiological differences in size are most prominent in animals that have to maintain a constant high body temperature (**homeothermic** animals—mostly mammals and birds). Comparative analyses show a correlation between body size and several physiological parameters, colloquially known as "mouse to elephant curves." These include heart rate (faster in small animals), longevity (greater in large animals), metabolic rate (higher in small animals), and others. All these are direct or indirect consequences of the relationship between surface area and volume.

As we can see, size has many effects on organismic form and function that reach far beyond basic geometry. We've already met some effects of size, without going into details. With an understanding of the effects of size, we can explain some phenomena we've already come across.

Larger animals have an increased need for transport systems, since oxygen and nutrition cannot reach all the tissues through diffusion. We will discuss this in greater detail in Chapter 24. Small aquatic animals can rely on surface breathing, whereas large animals need external gills. We already saw this in the diversity of marine annelids in Chapter 17. Small animals have less need for support structures, which can explain why many small-bodied phyla have secondarily lost the coelom or reduced the skeletal system. Larger animals need

Figure 19.1 Allometric size differences: Three herbivorous mammals, all drawn at the same size. Despite the fact that in the illustration, all these animals seem to be the same size, there are obvious differences in their proportions, stemming from the different physical constraints that come with each size. The largest animal (the elephant on the left) has relatively thicker legs and more robust neck muscles than the smallest animal (the deer on the right).

more complex digestive systems, since this system is essentially linear, but the animal's nutritional needs are based on its mass.

Indeed, most aspects of an animal's body plan are either strongly affected by size or constrain the minimum and maximum size the animal can attain. Animals with inefficient transport systems cannot attain very large sizes, whereas animals with high metabolic rates cannot be very small. Conversely, the relative proportions of different structures or appendages are affected by the animals' absolute size (Figure 19.1). When different structures change in size at the same rate as the organism increases in size, this is referred to as **isometry**. However, it is much more common for structures to change in size at different rates, a phenomenon known as **allometry**.

The evolution of size

In many lineages, most notably in vertebrates, there is a trend toward size increase over evolutionary time—a tendency known as Cope's Rule, after the nineteenth century American paleontologist Edward Cope. Larger animals are better able to avoid predation, and usually have to invest relatively less energy in maintaining physiological stability. However, increased size has its costs. There is often an upper limit to the size an animal can achieve, dictated by mechanical and physiological factors. The environment can support fewer individuals of large size, so population sizes of large animals tend to be smaller. Because of their increased dietary needs, large animals also tend to be more sensitive to changes in the availability of resources and are more susceptible to environmental crises. The combination of small population sizes and higher susceptibility to crises means that large animals are often the first to go extinct when there are changes in the environment. The trend of lineages to increase in size is counteracted by the vulnerability of larger animals, which is why we don't see all lineages reaching maximum size over time. Perhaps the best-known example of this trend of size increase and increased susceptibility to extinction is the dinosaurs. Many lineages within the dinosaurs grew to huge sizes, but all dinosaurs, with the exception of the lineage of small, feathered dinosaurs that led to modern birds, went extinct at the end of the Cretaceous.

On the other hand, we also see a tendency in some lineages to become smaller. Small size allows the utilization of resources and habitats that are unavailable to large animals, such as living in **interstitial** habitats (between the grains of the substrate, such as sand or soil). All the factors we listed as disadvantages to large size are positive factors for small organisms; larger population sizes, lower susceptibility to fluctuations in resource availability and more.

Different habitats exert selection for different optimal sizes. A well-known phenomenon is that homeothermic animals at higher latitudes (closer to the poles) tend to be larger than related animals at lower latitudes (such as arctic wolves vs. wolves in temperate climates), because the reduced surface to volume ratio of large animals allows for more efficient thermoregulation. Living on islands tends to have some surprising consequences, due a combination of factors both encouraging and increase in size (lack of predators) and a decrease in size (limited resources). Vertebrates on islands tend to converge on a similar medium size. Thus, we have both island gigantism of normally smaller taxa (e.g., tortoises) and island dwarfism of normally larger taxa (e.g., sheep-sized fossil elephants on Mediterranean islands).

Is there a measure for complexity?

There are many similarities between discussing the evolution of complexity and discussing the evolution of size. However, while size is an easily measured parameter, complexity is more difficult to define and to quantify. We can measure complexity in different ways. Morphological complexity can be defined as the number of distinct organs or systems or as the degree of elaboration of a given organ. However, these are very subjective measures and tend to be biased by the researchers' background and questions. A more objective measure for complexity is the number of cell types. When we introduced sponges, we pointed out that they are considered to be simple, because of the low number of cell-types they have. This measure is also problematic, both because the boundaries between cell types are sometime difficult to define, leading to different workers counting different numbers of cell types, and because the number of cell types is only one possible measure for complexity. It is possible to be morphologically complex, even with a small number of cell types.

We tend to compare the complexity of other animals to our own human level of complexity. Since we as humans have very complex brains, we tend to think of brain complexity as a central component of general complexity and dismiss animals with less elaborate brains as "simple." Throughout most of human history, the living world was seen as being arranged in a *scala naturae* "the ladder of nature," with simple animals sitting on the lower rungs, and increasingly complex animals sitting on higher rungs, up to the top of the ladder, which houses the pinnacle of creation—humankind (many depictions had higher rungs including angels, with God sitting at the highest rung). However, a research community composed of sponges might look at us with disdain due to the low number of organic compounds we are able to synthesize. Textbooks aimed at parasitic flatworms might list mammals among the simplest organisms because of our simple life cycles, including only a single stage.

The bottom line of all of this is that complexity can take many forms, and there is no single "axis of complexity." Over the years, there have been different philosophical schools regarding the evolution of complexity. One school sees complexity as a subjective and biased parameter. According to this point of view, there aren't more complex or less complex animals, but simply animals that have adopted different strategies for evolutionary success, and the question of complexity is irrelevant for discussing animal evolution. The opposite view sees complexity as measurable and important, and looks for objective measures of complexity, such as the number of cell types, genomic complexity, and others.

Does complexity increase throughout evolution?

If, for the sake of the argument, we adopt a position that is closer to the latter view that looks for objective measures of complexity, we can try to map complexity on the diversity of animals, both living and extinct. We can then ask if there is a trend for increased complexity in evolution (see Figure 19.2 for a hypothetical example). The picture that emerges is far from straightforward. It is clear that ecosystem complexity increased in the early stages of animal evolution, with several notable leaps in complexity, for example, the Cambrian Explosion (Box 11.1) or the Ordovician Biodiversification Event (Box 19.1). It is also clear that members of many specific lineages are more complex today—regardless of what parameter we choose to quantify this—than their earliest known ancestors. However, we have already seen many examples of secondary loss of complexity, for example, the body plan of Platyhelminthes and of several lineages within Annelida. The most extreme examples of secondary simplification we have seen are those common in certain parasites. Recall that parasitism has evolved numerous times and in almost every phylum. A common trend in the evolution of parasitic lifestyles is the loss of many complex structures relative to the ancestral condition. However, this is often accompanied by the increase in complexity of other characters.

We can say that over time, looking broadly at all animals, there is a general tendency for an increase in complexity within many lineages. However, this increase is offset by many cases of decrease in complexity, leading to a mix of animals with different levels of complexity that cannot be neatly arranged on anything like a *scala naturae*. We can metaphorically liken the evolution

of complexity to the popular board game "snakes and ladders" ("ropes and ladders" in some countries). We can compare increasing complexity to the player's advance from square to square, aiming to reach the top of the game board. The players' progress is sometimes aided by ladders that push them up the board (dramatic increases in complexity, in the metaphor) or hindered by snakes that send them down (decreases in complexity). However, the net result of all these ups and downs is a slow and constant progress upward.

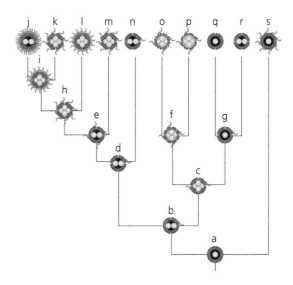

Figure 19.2 The evolution of complexity: a phylogenetic tree mapping the evolution of different elements of complexity in a series of hypothetical organisms. The organisms vary in their level of complexity in three traits: circles inside the body, integument, and appendages. Each of these traits can increase or decrease in complexity throughout evolution. The common ancestor of all the organisms on the tree (a) was simple in all three traits. The complexity of the appendages increases consistently in the lineage leading up to (j). However, in the lineage leading to (q) appendage complexity increases and then decreases. The complexity of circles increases consistently in the lineage leading to (p), but decreases in the lineage leading to (r). Species (s) is a "conservative" species, maintaining simple structures inherited from a distant ancestor. Nonetheless, it has relatively complex appendages. Species (q) is also relatively simple, but the phylogeny indicates that its simplicity is secondary, since it had more complex ancestors. Would you classify the lineage descended from (h) as more complex or less complex than the lineage descended from (f)?

Box 19.1 The Ordovician Biodiversification Event

In Box 15.1 we introduced the Cambrian Explosion as a key event in the evolution of complex animal life as we know it today. However, the world of the Cambrian was still very different from the world of the present, and the diversity of animals and of the ecological niches they occupied was still fairly low. It was only in the mid-Ordovician, roughly 470–450 million years ago, that diversity increased to the levels seen from then onward. This period of increasing diversity is known as the Ordovician Biodiversification Event (or sometimes, the *Great* Ordovician Biodiversification Event—GOBE).

The Ordovician Biodiversification Event saw the appearance of many animal groups that were to be dominant for the remainder of the Paleozoic and even later. Animals such as nautiloids and ammonites (both cephalopod mollusks—see Chapter 15), gastropods and bivalves (also mollusks), brachiopods (see Box 17.1), sea scorpions (marine chelicerates—Chapter 21), some of the earliest vertebrate groups (see Chapter 30), and many others appeared or increased in diversity. The number of Linnaean orders tripled: an increase seen in almost all phyla. The complexity of food webs increased, with the addition of intermediate ecological levels and increased specialization. In addition, the Biodiversification Event saw the expansion of life into deeper regions of the ocean and into open water, an increase in the diversity of reef communities, and the establishment of the planktonic realm.

While the Cambrian Explosion is far more famous, the Ordovician Biodiversification Event is probably the most significant event in the early evolution of animal diversity. The levels of diversity attained throughout the Ordovician remained until the cataclysmic extinction at the end of the Permian, which we will introduce in Box 26.1.

CHAPTER 20

Molting animals

Limitations of external skeletons

In Chapter 16 we discussed the dual role of external skeletons in support as well as protection. However, these benefits come at a price. The rigidity of the integument that functions as an external skeleton imposes limitations on the ability of the animal to grow, limits the range of movements possible with it and creates a strong separation from the environment. Some animals with external skeletons, such as the nematodes discussed later in this chapter, are able to move by flexing the skeleton laterally, because the skeleton has a certain amount of flexibility. In others, such as the arthropod discussed in Chapters 21 and 23, most of the skeleton is composed of hardened cuticle, to the point that it has almost no flexibility. In these animals, the skeleton is composed of individual plates that move relative to one another via joints.

Animals with rigid or semi-rigid external skeletons cannot increase in size within the skeleton. In order to grow they must remove and replace the skeleton with a new one of a larger size. Throughout evolution, animals with external skeletons have developed ways to remove and replace their external skeletons in order to deal with this requirement.

Strong separation from the environment has its advantages, especially if the environment is harsh or includes potential predators. However, as we saw in Chapters 7 and 12 the integument has an important role in providing information about the environment. A rigid cuticular external skeleton that protects against the environment also prevents information about the environment from passing through it. Animals with external skeletons have to evolve new sense organs that protrude from the skeleton or make use of gaps in the skeleton in order to gather information about their surroundings.

Growth through molting

The need to replace the external skeleton before growing leads to animals with external skeletons having a nonlinear mode of growth. Rather than growing continuously throughout their lives, they grow in short spurts wherein the animal sheds the skeleton, grows rapidly, and forms a new skeleton. The process in which the animal emerges from the old external skeleton is known as **molting** or **ecdysis** (Figure 20.1). Molting is followed by the secretion of exoskeletal components that harden to form the new skeleton. The stages between molts are known as **instars.** In a somewhat unfortunate choice of terminology, the instars are also sometimes referred to as molts.

The mode of molting, the number of instars, and the period between molts is highly variable. Some species will molt a fixed number of times throughout their life, whereas in other species the number of instars is variable. Some species have a terminal molt after which they reach maturity and do not continue to grow, whereas others will continue molting even as adults. Molting may involve a more or less dramatic change in morphology, or it may entail an increase in size without any morphological changes.

Molting is a complex, multistage process. It involves dissolving the connection between the external portions of the skeleton and the underlying integument, followed by breaking open the hard

Organismic Animal Biology. Ariel D. Chipman, Oxford University Press. © Ariel D. Chipman (2024). DOI: 10.1093/oso/9780192893581.003.0020

Figure 20.1 An illustration of an insect in the process of molting.

skeleton along specific lines of weakness or sutures. After the animal exits the old skeleton via these sutures, it increases in size, usually through filling dedicated spaces under the integument with air or water. It then has to synthesize the components of the exoskeleton, shuttle them to the correct location, and harden the new skeleton. The process of molting is mediated by a series of interacting hormones and other compounds, the exact identity of which varies among the different molting taxa. The process has been studied in the greatest detail in insects and crustaceans (see Chapter 23) and in nematodes (see later in this Chapter). The best-known regulators of molting are the compounds known as

juvenile hormone and ecdysone, but there are many others.

Ecdysozoa—the molting animals

Molting behavior of the type described above is concentrated in a single branch of the animal tree of life that is aptly named Ecdysozoa. Despite molting being a very clear defining character, Ecdysozoa was only recognized as a clade in the late 1990s. Its members were formerly allied with various other taxa and molting was not seen as a unifying character. The unification of the diverse phyla composing Ecdysozoa was a result of the advent of molecular phylogenetics, which attempted to use the newly available molecular techniques to uncover the relationships among phyla. As already discussed when we introduced Spiralia, these analyses recovered three main branches within Bilateria. Ecdysozoa is the largest of these branches in terms of the number of species and includes eight recognized phyla. In addition to molting, ecdysozoans share a lack of a ciliated epithelium (presumably a consequence of having a cuticular skeleton covering the epithelium that secreted it), a tri-layered cuticle that has chitin as one of its constituents, and a mouth that is positioned at the anterior-most end of the body (although in many ecdysozoans it has secondarily shifted to a more posterior ventral position).

Ecdysozoa is in turn usually subdivided into three main branches, each containing several phyla:

Nematoida—Elongated worms, including two phyla, Nematoda and Nematomorpha, which will be discussed later in this chapter.

Panarthropoda—Arthropods and their relatives, characterized by a metameric or segmental organization of the body and repeating appendages. Includes Arthropoda, Onychophora and Tardigrada. We will devote two full chapters to Arthropoda (Chapters 21 and 23), while the other two phyla will be dealt with briefly in a text box (Box 21.1).

Scalidophora—An assemblage of three small phyla of mostly endobenthic animals, Priapulida, Loricifera and Kinorhyncha. They are discussed briefly in a text box in this chapter (Box 20.1). There is some debate over the validity of this clade.

Nematoda—the round worms

Nematodes are quintessential "worms." They have a simple unsegmented body that is round in cross section (hence the common name) and is elongated and thread like (hence the scientific name—"nema" means thread in Greek), usually tapering to a point at both ends. They have no appendages, and no externally visible sense organs. They are mostly microscopic, rarely achieving lengths of more than a millimeter or two, but there are several exceptions among internal parasites that reach a length of 10 centimeters and rarely even more (the longest nematodes reach up to 8 meters). The external morphology of nematodes is very similar among species. Indeed, it is nearly impossible to identify species without resorting to microstructures visible only under an electron microscope. However, the diversity of nematodes is manifested through ecological and biochemical specialization. Nematodes can be extremely specific in the environment they inhabit, in the hosts they parasitize, or in the resources they use.

Nematodes are ubiquitous. They are found as parasites of all multicellular organisms, in the soil, in fresh water, and in the sea. They are agents of numerous tropical diseases (e.g., filariasis) and significant crop pests of all commercial crops. While the diversity and evolution of nematodes as a group is very poorly known and understudied, a single species, *Caenorhabditis elegans* is among the best studied lab organisms. When most biologists refer to "the worm" they mean *C. elegans* (rarely with the full generic name). We devote a text box (Box 32.1) to the importance of this species and other models in biological research.

Nematode body organization

Nematode body organization is much simpler than most of the phyla we have discussed up to now (Figure 20.2). Before molecular phylogenetics placed nematodes firmly within Ecdysozoa, this simple body plan was assumed to be primitive.

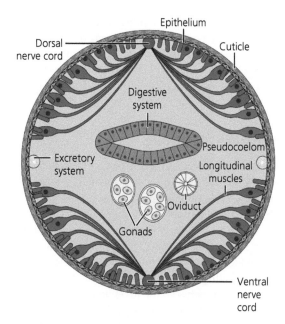

Figure 20.2 Transverse section through a generalized nematode.

Given their position, it is more likely that their simplicity is a result of secondary simplification (see Chapter 19), although it is not clear what their more complex ancestors would have looked like.

Like all ecdysozoans, nematodes are characterized by an external cuticle. This cuticle is multilayered, but is relatively flexible, allowing them to move without needing joints between individual plates. The cuticle is composed of various structural proteins strengthened by polysaccharides and lipids. The cuticle provides protection against many hostile chemicals and is resistant to water loss, allowing nematodes to live in challenging environments, ranging from the acidic gut of host animals to dry soil. Chitin is found only in the cuticle of the pharynx.

Nematodes do not have a complete coelom, and their body cavity—when they have one—is classified as a pseudocoelom (see Chapter 16). This means that the body cavity surrounds the endoderm of the digestive system without any mesoderm between the cavity and the gut. The body cavity fluids are under very high pressure, and this high-pressure fluid provides the main structural support. Nematodes are also unusual in only having a single layer of longitudinal body wall musculature, in contrast with the more common arrangement of longitudinal and circular muscle layers. This arrangement of muscles allows them a characteristic serpentine mode of locomotion either within the substrate, on the surface or in water.

Nematode life history

Nematodes as a rule reproduce sexually, with most species having separate males and females, some having females and hermaphrodites, and some having only hermaphrodites. Parthenogenesis is rare. Reproduction through fission or budding does not occur in any nematodes. The typical life history of nematodes includes four molts. Development is direct, with no clear morphological or ecological distinction between the juvenile stages and the adults aside from an increase in size. Some species of nematodes have a resting or **dauer larva** stage, in which the animal goes into a type of stasis where it is protected from the environment and slows down or stops physiological processes in order to survive periods of harsh conditions.

Embryonic development of nematodes is known mostly from *C. elegans*, where it has been studied in very great detail. The development of *C. elegans* has been mapped at the level of every individual cell through the entire developmental process. Indeed, the number of cells in the *C. elegans* body is fixed at every instar, and the process of development is conserved and stereotypical.

Feeding and digestive system

As already mentioned, the diversity of nematodes lies mostly in their specific adaptation to different environments and feeding modes. Bacteria and protists are the most common source of nutrition for nematodes. These single-celled organisms usually live on the substrate that the nematodes inhabit. Thus, nematodes are often found on or in decaying organic matter, feeding on the organisms that drive the decay. Other nematodes feed directly on decayed matter within the substrate. Parasitic nematodes are common and probably compose the

majority of nematode diversity. Nematode parasites often feed directly on the host's body fluids or on dissolved nutrients. When they become numerous, they can block circulatory vessels, causing damage to the host. This is the case in filarial nematodes that block the lymphatic system of human hosts, causing buildup of liquids in connective tissues (sometimes referred to as "elephantiasis"). Some parasitic nematodes are carriers of bacteria that attack and break down the tissues of the host, providing nourishment for the parasite.

Regardless of the feeding mode, nematodes have a simple linear digestive system, mostly composed of a single layer of epithelial cells. The mouth opening is at the anterior end of the body. The pharynx is muscular and is lined with hard, chitin-containing cuticle. The shape of the pharynx is typically roughly triangular in cross section and is known as a **tri-radiate pharynx**, a structure common to several other ecdysozoans phyla. Digestive glands are found in the pharynx and along the digestive system. Toward the end of the digestive system is a cloaca that reabsorbs excess water. The digestive system terminates in an anus at the posterior tip of the body.

Nervous and sensory systems

As with other organ systems, the nematode nervous and sensory systems are simple. The anterior nervous system is a ring surrounding the esophagus. This ring connects to a ventral nerve cord with several ganglia. A dorsal nerve tract is composed of axons only and does not include cell bodies. All in all, the nematode nervous system is composed of only a few hundred nerve cells. In some species, the nervous system is known at cellular resolution. It is clear from the study of these systems that the number and arrangement of neurons in all individuals of a species is identical. Differences among species are fairly small.

Nematodes have no photoreceptors, and the main sources of sensory input are chemoreception and touch. Sensory receptor cells are concentrated in anterior sense organs.

Nematode diversity and evolution

While only a few tens of thousands of species are formally identified and described, the number of actual species on the planet is almost definitely much higher, with some estimates talking about tens or hundreds of millions of species. Nematodes are found everywhere—in the water column and on the bottom in both sea and fresh water, in terrestrial soil, on and in plant leaves and roots, and on and in almost all animals on the planet. The largest identified diversity of free-living nematodes is in the sea. However, the world of parasitic nematodes is significantly underexplored. There are claims that every species on the planet has its own species-specific parasitic nematodes. If this claim is true, the number of parasitic species only would exceed the total number of non-nematode species in the world.

Nematomorpha—the horsehair worms

Based on both morphological characters and on molecular phylogenetics, the nematomorphs are the closest sister group to nematodes. They are superficially similar to nematodes, but tend to be much longer and thinner, resembling a long thin thread or hair. They are usually 5–10 cm long, with the longest species reaching up to 1 meter in length. All nematomorphs are internal parasites—usually of insects or other arthropods. They have a biphasic life history, with the juvenile stages inhabiting the body cavity of their host and consuming it from within, and the adults living in fresh water (rarely in the sea). The parasitized host will normally seek out a water source and drown there, allowing the adults to emerge by rupturing the body wall. The adults have a degenerated digestive system and do not feed at all, living only long enough to find a mate and reproduce.

Box 20.1 Scalidophora

The three phyla that make up Scalidophora are composed of marine benthic or endobenthic worms and have relatively low species numbers. Scalidophora is thus by far the smallest and least-known of the three ecdysozoan clades. The clade name means "scalid-bearing," **scalids** being typical small spines that are common, and probably unique, to the three phyla. In addition to scalids, scalidophorans share a similarly shaped eversible head structure, known as an **introvert**.

Priapulida (penis worms)—Only 22 species of priapulids exist today. They are fat, worm-like animals, their name (for the Greek god Priapus) alluding to a shape slightly reminiscent of human male genitalia. The smallest species are about 1mm in length, whereas the largest reach up to 40cm. Priapulids burrow in the sea floor and hunt small invertebrates using their introvert. Priapulids were an important component of the benthic community in the Cambrian, and more species of Cambrian priapulids are known than extant ones.

Kinorhyncha (mud dragons)—Microscopic animals that normally live in the sediment or attached to sessile, bottom-dwelling animals. Over 250 species are known, all under 1mm in size. Kinorhynchs have an externally segmented body and a flexible, retractable mouth cone, which gives them their name (literally "movable nose").

Loricifera—The first species of Loricifera was described in the 1980s, with another 37 additional species described since then. Loriciferans are another phylum of microscopic animals that live within the sediment. As we have seen, these phyla tend to be poorly studied and not very speciose, but also tend to have surprising shapes and body plans (at least when compared with the more commonly encountered phyla). Loriciferans have a prominent scalid-bearing introvert and their body is surrounded by a corset-like covering of robust plates ("lorica" is Latin for corset). They have complex life cycles, including several life stages, and alternate developmental paths.

CHAPTER 21

Arthropoda I

General introduction and Chelicerata

The arthropods

The arthropods are among the most familiar and most common animals on earth. In fact, it is likely that the majority of animals we see on a day-to-day basis are arthropods. The arthropods include insects, crabs and crayfish, spiders, ticks and mites, millipedes and centipedes, and a host of other better or less well-known animals. By any metric, the arthropods are the most successful group of animals on the planet (with the possible exception of nematodes—see Chapter 20). There are more arthropod species described than all other eukaryotes combined—over 1.2 million in all. Arthropods are found in almost all environments, except for the most extreme polar regions. They feed on almost all possible food sources. They are the ecologically dominant species in rain forests, in the air, in the deep sea, and elsewhere. They range in size from microscopic mites of less than 100 μm to giant crabs with a leg-span of over a meter from tip to tip.

Despite this diversity, arthropods are all variations on a conserved body plan. They all have a segmented body with segmented limbs, and the different segments and different limbs can be modified for specific functions, giving rise to the myriad morphologies found in the group.

Arthropods are traditionally divided into four Linnaean classes (but some sources consider them to be subphyla): Chelicerata—spiders, scorpions, and their kin; Hexapoda—insects and related taxa; Crustacea*—crabs, shrimp, lobsters, and other mostly aquatic taxa; and Myriapoda—centipedes and millipedes. In a phylogenetic framework, crustaceans form a paraphyletic taxon (hence the

asterisk following their name—see Chapter 4), since hexapods emerged from within them. The earliest branching event in arthropod phylogeny is between Chelicerata on the one hand, and Mandibulata on the other, with Mandibulata encompassing the remaining three classes.

In this chapter we will introduce the general characteristics of arthropods as a phylum. We will then describe the first branch of the arthropod tree, the chelicerates. Chapter 23 covers the second and larger branch, the mandibulates. Being so diverse, the arthropods cannot be covered in as much detail as they deserve in the space available in this book. We will give the generalities, but there is of course much more to arthropods than generalities. Some specific examples will serve to highlight the range of adaptations and specialization that can be found in this fascinating phylum.

Arthropod body organization

As already mentioned, the universal defining feature of the arthropod body is a segmental organization. All arthropods are made up of a series of segments along the anterior–posterior body axis. The number of segments is rarely below 10 and can be as high as several hundred segments in some millipedes. The ancestral state is for each segment to have a single pair of limbs or **appendages**, as seen in centipedes. However, in most arthropods, at least some of the segments have lost their appendages. Arthropod limbs are themselves segmented (or jointed), each being made up of a number of units that can bend around joints. Indeed, the

Organismic Animal Biology. Ariel D. Chipman, Oxford University Press. © Ariel D. Chipman (2024). DOI: 10.1093/oso/9780192893581.003.0021

name "Arthropoda" means having jointed limbs. This is in contrast to the arthropods' close relatives, the onychophorans and tardigrades (see Box 21.1) who also have repeated limbs along the body axis, but these limbs are not segmented.

All arthropods have a hard chitin-based exoskeleton. In some taxa, the cuticle forming the exoskeleton is strengthened by the addition of minerals. For example, in crustaceans, the cuticle is strengthened by the addition of calcium. Some insects that feed on tough food items have feeding organs hardened by heavy metals. The exoskeleton is composed of individual plates. The dorsal plates are known as **tergites**. The ventral plates are known as **sternites**. The lateral plates connecting the two, when present, are known as **pleurites**. Adjacent plates are usually connected by a flexible membrane, which allows movement of the joints. The jointed limbs articulate with the ventral or ventro-lateral side of the body. The limbs are protected by tubular skeletal elements.

Arthropods have a clear anterior region that includes the brain, major sense organs, and the mouth, which faces ventrally. In insects and myriapods, this region is enclosed in a separate and distinct **head capsule**. In chelicerates and in most crustaceans, this region is fused with a region that also includes walking limbs, to give a **prosoma** or **cephalothorax** (see Figure 21.1 and details in the description of the individual classes).

Arthropod segments are not all identical. Some have appendages adapted for locomotion, and some have appendages adapted to feeding or prey capture. Some segments bear appendages with highly specialized functions—such as web weaving in spiders—while others do not bear appendages at all. Segments bearing appendages with similar function tend to be clustered together in distinct body regions known as **tagmata** (singular **tagma**). The organization of the tagmata is known as **tagmatization**. This is specific for each of the four classes (Figure 21.1).

Arthropod life history

Arthropods reproduce sexually in almost all cases. The sexes are separate and fertilization is internal. In most arthropod species, mating is direct, and the reproductive organs of the male and female fit together in a "lock and key" mechanism. In some taxa, mating is via a chitinous sperm-containing package, known as a **spermatophore**, which the male either leaves for the female to pick up, or delivers directly to the female's reproductive opening. Hermaphrodity is very rare and is only found in a few species of barnacles—sessile marine crustaceans. Parthenogenesis is also found in a few taxa, most notably in aphids, which at least in some stages of their life cycle have only females, with production of offspring taking place without the involvement of a male.

Growth in arthropods is discontinuous through distinct molts as in all Ecdysozoa. The individual inter-molt stages are known as instars, or stadia depending on the specific taxon. In many cases, there is an adult terminal molt, after which the animal no longer grows. This is true for all insects and for a few other taxa. In some cases, there are post-adult molts and the animal continues to grow and molt throughout its life. Both direct developing and biphasic development are found among arthropods. The transition between juvenile or larval stages (known as **nymphs** in some taxa) and the adult form sometimes includes a dramatic metamorphosis. This is best known in holometabolous insects (see Chapter 23). Many marine crustaceans have larval stages that are morphologically and ecologically different from the adults, but the transition between them is gradual and not dramatic. Several groups of decapod crustaceans have a series of larval stages, each with its own adaptations.

The arthropod body plan is mostly laid down during embryonic development, and the hatchlings of many arthropods already have all or most of their adult segments. The situation in which all segments are formed during embryogenesis is known as **epimorphic** development, whereas the situation where segments are added in successive postembryonic instars is known as **anamorphic** development. Examples of both types of development are scattered throughout arthropod phylogeny, including many variations on them. Insects are all epimorphic, as are most chelicerates (anamorphic chelicerates include mites, ticks, sea spiders, and horseshoe crabs). Millipedes, some centipedes, and many

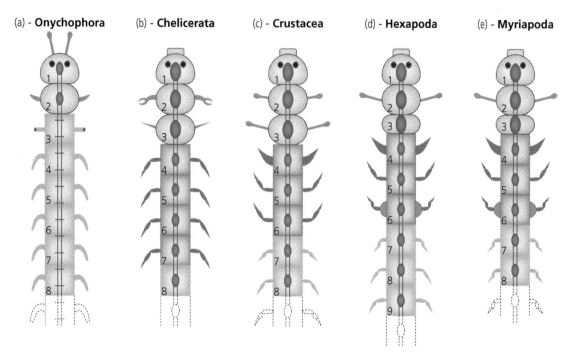

Figure 21.1 Organization of segments into tagmata in the different arthropod classes: A schematic representation of the identity of different segments in the four arthropod classes, and in their closest relatives. Brain segments (pre-gnathal segments) in orange, head-related segments in pink, trunk segments in green. The nervous system is in blue, with the brain ganglia illustrated larger than trunk ganglia. Segments in the same position in different classes are believed to be homologous. (a) In Onychophora, the arthropods' closest sister-group, The head is composed of two segments. The first segment 1 bears a pair of antennae and a pair of eyes. Segment 2 bears the jaws. Segment 3 has unique slime papillae. All segments from 4 onward (up to approximately 30 segments) bear a pair of un-arthropodized limbs. (b) The first segment in chelicerates bears eyes and a structure known as the labrum, which may be homologous to onychophoran antennae. Segment 2 bears feeding appendages known as chelicerae. Segment 3 bears sensory appendages known as pedipalps. Segments 4–7 each carry a pair of walking limbs. Segments 1–7 together form the prosoma. The segments from 8 and onward bear no walking appendages, although some may have a range of specialized appendages. These posterior segments together form the opisthosoma. (c) In crustaceans, the first segment bears a pair of eyes and the labrum. Segments 2 and 3 each carry a pair of antennae. Segments 4–6 carry feeding appendages or gnathal appendages. Segments 1–6 together form the crustacean head, but they are not always enclosed in a head capsule. Segments 7 and onward are highly variable in an order- or family-specific manner. They may carry accessory feeding appendages, walking limbs, swimming limbs, gas exchange organs, and others. (d) In Hexapods, the first segment bears a pair of eyes and the labrum, and in some cases unpaired ocelli. Segment 2 bears the antennae. Segment 3, the intercalary segment, is reduced and carries no appendages. Segments 4–6 carry feeding appendages. Segments 1–6 together form the insect head, which is enclosed in a distinct capsule. Segments 7–9 form the thorax, with each segment carrying a pair of walking limbs. In most insects, segments 8 and 9 also carry wings. Segments 10 and onward (usually a total of 9–11 segments) form the abdomen and bear no appendages.
(e) The anterior six segments of myriapods are identical in organization to those of hexapods. Segments 9 and onward form the trunk and bear a pair of appendages per segment. In centipedes there can be between 15 and up to close to 200 trunk segments, with the anterior trunk appendage modified as a venom delivery organ. Millipedes have up to several hundred segments, with each pair of segments usually fused to give a diplosegment.

crustaceans are anamorphic developers. The most extreme anamorphic development is found in many crustaceans and sea spiders, which hatch with only four segments into larval forms known as a **nauplius** (crustaceans) or a **protonymphon** (sea spiders) and add segments throughout a series of larval stages. The most extreme epimorphic development is found in geophilomorph centipedes, which hatch with dozens of segments and do not add any post-embryonically.

Feeding and digestive systems

There is a huge diversity of feeding modes within arthropods. For almost any possible source of nourishment, we can find arthropod species that utilize it. Barnacles and some planktonic crustaceans are suspension feeders. Most chelicerates and nearly all centipedes are predators. Many species of insects are herbivores. Millipedes are herbivores or detritivores. There is a diversity of insects that live exclusively on a liquid diet, either throughout their lives or as adults only. This includes nectar-feeding butterflies, blood-feeding mosquitoes, sap-feeding bugs, and many others. Termites feed on difficult-to-digest wood, making use of endosymbionts to break down lignin, the main component of wood.

Anterior appendages of arthropods are modified as feeding organs. These appendages have evolved to fit the specific requirements of the species' diet and feeding mode. Mandibulate arthropods usually have a series of three feeding appendages, the anterior one being a **mandible**, which gives the group its name. Mandibles are primitively biting or cutting structures, but they have been modified for other purposes in many taxa. The mandible is followed by two pairs of appendages, usually known as **maxillae**. The second pair in insects and in myriapods is fused to give a single plate known as the **labium**.

Chelicerates have anterior feeding appendages known as **chelicerae**. These are usually piercing or cutting appendages and are rarely used for processing food. Most chelicerates pierce their prey and inject digestive enzymes into it, sucking up the digested material. The chelicerae are followed by a pair of **pedipalps**, which are modified for a range of uses in different orders, which may or may not be related to feeding, as detailed below. An anterior unpaired appendage, the **labrum** is found on the first segment of all arthropods. Its role varies, but it often forms an upper protective covering to the mouth.

The digestive system is normally straight, without a coiled intestine. It starts in a ventral mouth and ends in a posterior anus. The gut is divided into three sections, the foregut, midgut, and hindgut. The foregut and the hindgut are lined with cuticle. Endosymbionts are very common in arthropods and aid in digesting different food types,

as already mentioned in termites. These endosymbionts, which may be bacteria or protists, are often found in blind extensions to the gut known as **caeca** (Figure 21.2).

Circulatory system and gas exchange

The circulatory system in arthropods is of a type known as an open circulatory system (see Chapter 24). The circulatory fluid is called **hemolymph** and it is mostly found within a modified coelomic cavity, the **hemocoel**. Arthropods have a dorsal muscular heart that pumps hemolymph anteriorly. The hemolymph is collected from the hemocoel through a series of segmental openings or **ostia** in the heart. The hemolymph then empties from the anterior of the heart back into the hemocoel, maintaining a constant flow. The hemocoel is dorso-ventrally divided into three cavities, separated from each other by membranes. The dorsal cavity contains the heart while the ventral cavity contains the nervous system. The intermediate cavity is the largest and contains the digestive and reproductive systems.

The hemolymph serves multiple functions. It is the main transport system for the products of the digestive system and for oxygen (but not in insects). It is the site for most immune functions, and it acts to seal wounds. It is also a signaling medium for various hormones. The hemolymph contains cells, mostly immune cells. There are no specialized oxygen-carrying cells. It is normally unpigmented, but in some large crustaceans and chelicerates it contains hemocyanin, which acts as a soluble oxygen carrier.

Different arthropods display diverse modes of breathing and gas exchange. Many crustaceans use modified limbs as gills, whereas some of the smallest crustaceans exchange gas directly through the body wall. Horseshoe crabs use specialized multilayered structures in the **opisthosoma**—**book gills**—for gas exchange. Many terrestrial arachnids use a modified version of these, known as **book lungs**, for aerial gas exchange. Insects and some arachnids use **tracheae**, branching tubes that connect to the outside air for delivering oxygen to their tissues.

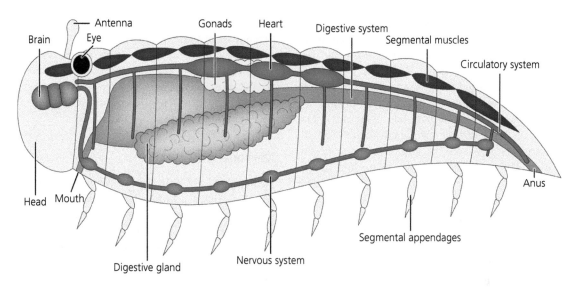

Figure 21.2 Organization of the body in a generalized arthropod: Lateral view of the main organ systems. The illustration includes elements from different arthropod classes and is not meant to represent any one class. The structure of the head is not elaborated.

Muscles and movement

The structure supporting movement in arthropods is the external skeleton. Arthropods have a variety of muscle types. These muscles normally connect to inwards-facing processes in the skeleton. Longitudinal muscles connect adjacent segmental plates, allowing movement of segments relative to one another. Dorso-ventral muscles connect the tergites to the sternites of the same segment (Figure 21.3). In winged insects, these muscles provide most of the force for flight. Limb muscles include both external muscles that connect the limb cuticle to the main body wall and internal muscles that connect adjacent segments of the limb. The exact number and location of limb muscles varies depending on the types of movement executed by the limb.

The primitive locomotory mode of arthropods was probably walking along the sea floor, and the appendages originally evolved as locomotory structures. Walking remains the most common form of locomotion in arthropods, although most extant species are terrestrial. The number of walking limbs varies from 6 in all insects, through 8 in almost all chelicerates, roughly 10–20 in crustaceans, and up to tens of legs in centipedes and hundreds in millipedes (Figure 21.1). Locomotion is through tightly coordinated movements of the limbs, with

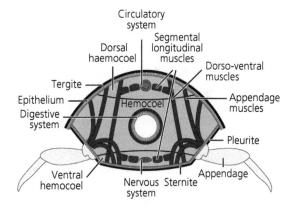

Figure 21.3 Transverse section through a generalized arthropod.

different arthropods having different gaits. Alternating lateral leg movements is the norm in insects, whereas multi-legged millipedes exhibit waves of leg movement running symmetrically down the body. Even in long-bodied arthropods, there is usually no body flexion involved in locomotion, and body wall musculature is not involved in locomotion. This is in contrast with long-bodied annelids or nematodes, which move primarily through serpentine locomotion.

Most species of insects are capable of powered flight. Wings are extensions of the external body wall, connected to the body by compound joints. There are normally two pairs of wings, although many insect taxa have secondarily lost one pair (e.g., flies and mosquitoes). Wings have been lost completely or fused to the tergites in several insect taxa either in all stages (e.g., many beetles) or in specific stages or individuals (e.g., ants). The mode of flight varies considerably depending on the insect's size and the size and shape of the wings. The wings not only move up and down, but also change their angles and move forward and backward. These complex movements are delicately coordinated through a series of flight muscles at the base of the wing and the dorso-ventral body muscles.

Another unusual mode of aerial locomotion is ballooning, which is found in many species of spiders (and in a handful of insects). In ballooning, the spider—usually a hatchling—releases a long thread of silk, which catches the wind and lifts the spider, transporting it with the air currents. Ballooning is an important form of dispersal in the species that practice it, transporting the hatchlings to distances of up to hundreds of kilometers.

Swimming is found in many groups of crustaceans as well as in a small number of aquatic insects, and in horseshoe crabs and sea spiders (aquatic chelicerates). In swimming arthropods, some of the limbs are modified to form paddles. These paddles are used for rowing or steering and are coordinated in the same way walking limbs are. In some very small crustaceans, the swimming appendages are not paddle-like but feathery. This is a result of the unusual characteristics of fluid dynamics at very small scales (see Chapter 19).

Arthropod limbs are not always built on a single axis. Crustacean limbs have a double axis and are referred to as **biramous** limbs. This is in contrast with the limbs of most other arthropods that are **uniramous**. The main axis of the biramous limb is known as the **endopod** and is usually used for locomotion. The secondary axis is the **exopod** and this is often modified as a gill structure.

While most arthropod locomotion is achieved by movement of the appendages, it is worth mentioning some cases where it is movement of body segments relative to each other that drives locomotion. This is most evident in predator evasion responses

that involve a "flick" of the body through the sudden release of stored energy in muscles. This is seen in jumping of springtails and in the escape response of many crayfish and shrimp. Rapid release of stored muscle energy in the limbs is also what allows many insects to jump rapidly and suddenly, best known from grasshoppers and fleas.

Nervous system

Arthropods have a ladder-like central nervous system. It is composed of segmental ganglia with two nerve tracts or **commissures** linking adjacent ganglia. In many cases, two or more ganglia are fused, resulting in the presence of less than one ganglion per segment. However, even when segmental ganglia are fused, they can be seen as independent nascent units in embryonic development. The brain is formed of three anterior dorsal ganglia. These three ganglia are named (from anterior to posterior) the **protocerebrum**, **deutocerebrum,** and **tritocerebrum**. The segments that house these ganglia are accordingly named the protocerebral, deutocerebral, and tritocerebral segments (Figure 21.1).

As can be expected from their diverse morphology and ecology, there is a corresponding diversity in arthropod sensory organs. Both simple and compound sensory organs can be found. Many of these are concentrated on the head, but many are also found throughout the body, on appendages, and attached to the external body wall.

With few exceptions, arthropods have eyes as one of the sensory organs on the head. Several groups of arthropods have a pair of compound eyes, composed of numerous photoreceptive units known as **ommatidia**. Compound eyes are found in insects, some centipedes, horseshoe crabs, and some crustaceans, and this is probably the ancestral state for arthropods. In other taxa, the eyes are reduced to simple eyes made up of a single ommatidium and these are known as **ocelli**. In these cases, there are often more than two simple eyes. Median, unpaired ocelli are common. There are also combinations of simple and compound eyes. Many insects have a pair of compound eyes, a median ocellum, and often two additional lateral ocelli. Spiders have six or eight ocelli, and in some spider taxa one pair is large and forward facing.

Chemical reception or chemoreception is a significant part of the arthropods' sensory field. Chemosensory cells are found throughout the entire body, but they are mostly concentrated in the antennae or pedipalps, in feeding appendages, and in other appendages. Chemoreception has been widely studied in many arthropod species. Some species are known to be sensitive to specific compounds at low concentrations, notably female pheromones or indicators of specific food sources. Arthropods constantly taste their surroundings using their diverse sensory appendages, to find food, mates, and egg-laying sites. The ovipositor (egg-laying appendage) of grasshoppers is known to have chemosensory hairs that judge the suitability of the substrate in which the female is laying.

Mechanosensory organs are found in the form of bristles that are sensitive to pressure, vibration, or air and water movement. Indeed, the hairy appearance of many arthropods is due to the distribution of sensory hairs and bristles over the body. Mechanosensory cells are also present in the joints of the body and limbs, providing feedback about the position of the joints and the stress imposed on them, allowing better regulation of locomotion.

The **antennae** are long appendages at the front of the head of mandibulate arthropods. Crustaceans have two pairs of antennae, whereas insects and myriapods have only one pair (Figure 21.1). The antennae host a combination of chemosensory and mechanosensory organs and function as a major sensory structure. In chelicerates, who lack antennae, their role is mostly filled by the pedipalps or, in some groups, by greatly elongated walking legs. Many mandibulates have posteriorly facing sensory structures known as **cerci**. Like the antennae, these have a combined mechanosensory and chemosensory role, and are often specifically sensitive to air movement, alerting the animal to the possible approach of a predator from behind. Cerci also play an important sensory role during mating and egg-laying of many insect species.

Evolutionary history

Because of their hard and easily preserved exoskeletons, arthropods have an excellent fossil record. Fossils of arthropods and their close relatives are found from the Lower Cambrian and onward. Most of these fossils are of exoskeletal elements only, but there is an increasing number of fossil sites with exceptional preservation, in which the full body, including soft parts, is preserved. Early arthropods and their relatives are well represented in these sites. This detailed record allows the reconstruction of the early stages of the arthropods' evolutionary history. Lobopods are a diverse paraphyletic assemblage of fossils, representing body plans that are ancestral both to arthropods and to the other panarthropod phyla. They have annulated bodies with unjointed legs and simple anterior structures. Successive groups of fossils show the gradual acquisition of fully segmented bodies, jointed legs, compound heads, and other arthropod characteristics. These include many bizarre taxa such as the radiodonts—arthropod relatives with lateral flaps for swimming and giant anterior hunting appendages—and *Opabinia*, another swimming arthropod relative with five eyes and a single flexible trunk-like anterior appendage.

Members of Arthropoda are first represented in the fossil record by trilobites. The trilobites were a diverse group with nearly 20,000 described fossil species, which were among the dominant animals on Earth until their extinction at the End Permian extinction 250 million years ago (see Box 26.1). Trilobites had a hard, mineralized exoskeleton making them easily fossilizable. They were flattened animals, mostly substrate dwelling with three characteristic longitudinal lobes.

The arthropod fossil record indicates that by the Late Cambrian (about 500 million years ago), chelicerates, and mandibulates had already diverged. There is some debate as to whether trilobites belong within Mandibulata or not. If they are mandibulates, then the chelicerate–mandibulate split would have to be even earlier, before the Middle Cambrian (more than 520 million years ago). Representatives of the modern classes are already known from the Silurian (about 440 million years ago), with the exception of insects, which were relatively late arrivals and are known only from the Devonian (about 400 million years ago).

Chelicerates

We now turn to take a closer look at the first branch of the arthropod tree, the chelicerates. The chelicerate body is made of two tagmata known as the **prosoma** and **opisthosoma** (in some sources the terms used are cephalothorax and abdomen, but these terms are confusing since they are also used for non-homologous tagmata in mandibulates). The prosoma includes four segments with locomotory appendages, and two anterior segments with specialized appendages, the chelicerae and the pedipalps, already mentioned above. In the very anterior is the **ocular** segment, which bears the eyes. The prosoma thus combines locomotory, sensory, and feeding functions. The opisthosomal segments have either no appendages or only small appendages with specialized functions. The chelicera is an unusual appendage, originally including an elbow-like hinge and terminal pincers. However, the pincers are missing in some lineages. Although the chelicera is primarily a feeding organ, it is very different from mandibulate feeding organs and is usually incapable of biting or chewing. The majority of chelicerates are predators or blood suckers, and at least four groups have evolved venom.

Chelicerate diversity

Chelicerata includes two marine taxa, Pycnogonida and Xiphosura and a diverse terrestrial taxon, Arachnida. The pycnogonids or sea spiders are a small group of arthropods with some 1300 described species. They have an unusual body plan, with a small and thin body, spindly legs, and a highly reduced opisthosoma. Due to the lack of an abdomen-like region, and because the body is so thin, major organ systems extend into the legs. Most pycnogonids have seven pairs of appendages. The anterior includes two pairs of appendages called **chelifores** and **palps**, which are likely homologous to chelicerae and pedipalps respectively. A unique paired appendage found only in sea spiders is the **oviger**, which is used by males to carry eggs. The majority of pycnogonids then exhibit four pairs of walking legs, though five and six pairs have evolved independently in some groups. At the anterior end is a proboscis used for feeding. Simple eyes

are found on an eye tubercle that rises above the prosoma. Pycnogonids are not well studied, and there are many open questions about their biology. They are generally accepted to be the sister group to all other chelicerates, and their body plan is highly modified relative to the ancestral chelicerate one.

The xiphosurans or horseshoe crabs (not crabs at all!) contain only four living species. However, they have a long evolutionary history with many fossil taxa known dating back 450 million years. The horseshoe crabs are very conservative and have changed little throughout their evolutionary history. The horseshoe prosoma is covered by a thick dorsal carapace, with the flexible opisthosoma sticking out posteriorly from the carapace. They have an elongated posterior spine that gives the group its scientific name (literally "sword tails"). They can walk along the substrate or swim with their ventral side up, using their flattened paddle-like limbs. Horseshoe crabs are used in bio-medical research because of the useful properties of their blue-colored hemolymph. The hemolymph includes proteins that clump around lipopolysaccharide from gram-negative bacteria, and can therefore be used to test for the presence of these toxins in medical samples. Horseshoe crabs are also known for the mass breeding assemblages on the shore during high tide in the spring and early summer.

The arachnids are the terrestrial chelicerates and comprise the majority of chelicerate diversity. Over 130,000 species are divided into at least 14 orders. Arachnida is generally considered to be a monophyletic group, descended from a single terrestrialization event in the Ordovician, about 485 million years ago, or earlier. However, recent molecular phylogenies have recovered Xiphosura as nested within Arachnida, rendering the arachnids a paraphyletic group, and suggesting more than one terrestrialization event. This result is not broadly accepted by researchers in the field, but if true, it would require a rethinking of chelicerate evolutionary history.

The main orders comprising Arachnida (Figure 21.4) are listed below:

Araneae (spiders)—The largest and most diverse arachnid group with over 50,000 species. Spiders are characterized by the presence of posterior silk spinning organs known as **spinnerets**.

(a)   (b)

(c)   (d)

Figure 21.4 Examples of chelicerate diversity: (a) A scorpion, the deathstalker *Leiurus quinquestriatus*. (b) A solifuge, the camel spider *Solpugena*. (c) An acarid, the predatory mite *Anystis*. (d) A jumping spider, *Hyllus semicupreus*

Source: a: Photo supplied from Shutterstock: AN Protasov (https://www.shutterstock.com/image-photo/palestine-yellow-scorpion-deathstalker-leiurus-quinquestriatus-579350458); b: Photo supplied from Shutterstock: Ondrej Michalek (https://www.shutterstock.com/image-photo/camel-spider-solpugema-sp-ndumo-game-1722892828); c: Photo supplied from Shutterstock: Tomasz Klejdysz (https://www.shutterstock.com/image-photo/anystidae-anystis-sp-predatory-mite-hunted-1125728078); d: Photo supplied from Shutterstock: Common Human (https://www.shutterstock.com/image-photo/macro-closeup-on-hyllus-semicupreus-jumping-1741445750).

Not all spiders weave webs to catch prey, but all spiders produce silk, and all spiders are predators. Spiders can be broadly classified into ground spiders and orb-weaving spiders. Spider pedipalps function as male secondary reproductive organs. The male collects his sperm and packages it in a specialized sperm-bearing structure in the pedipalps. The tip of the sperm-loaded pedipalps is inserted into the female's genital opening for fertilization. Spider chelicerae are connected to venom glands, and prey is immobilized by envenomation by the chelicerae.

Scorpiones (scorpions)—Characterized by an elongated and flexible opisthosoma (the scorpion's "tail") with a sting at the tip. Scorpions have potent venom, that in some species is dangerous (or even deadly) to humans. The scorpion's claws are enlarged pedipalps, which function in holding and processing prey and in mating. Mating follows an elaborate courtship dance. Fertilization is internal, with sperm delivered in a spermatophore. Development is also internal in most species, with live young hatching directly from the mother. Hatchlings are carried on the mother's back until they can care for themselves. Some 2500 scorpion species are known.

Pseudoscorpiones (false scorpions)—A group of tiny arachnids, with over 4000 species, usually no more than a few millimeters in size, which are an important component of soil ecosystems.

False scorpions have pedipalps that resemble scorpion claws but do not have an elongated opisthosoma. Venom glands are found inside the claws in most species and are used to envenomate prey. Silk glands are found in the chelicerae.

Acari (ticks and mites)—About 55,000 species of mites and ticks are known, but this is probably an underestimate. Acari are separated into two distinct orders Acariformes (mites) and Parasitiformes (ticks and others), but we treat them together here. They are characterized by fused opisthosoma and prosoma, covered by a shared carapace. Ticks are larger, reaching up to a centimeter in size and are usually blood-sucking ectoparasites. Mites are small to microscopic and are found in almost all environments. Microscopic dust mites and skin mites are ubiquitous in human habitation and are often the source of allergic reactions. Mites are significant pests of many crops as well as of bee colonies. Ticks are a public health risk both as parasites and as carriers of more severe diseases such as Lyme disease. Mites and ticks are unusual among arachnids in hatching with three leg pairs and adding the fourth pair in the first post-embryonic molt, a condition they share with Ricinulei, a small tropical arachnid order commonly known as hooded tick spiders.

Opiliones (harvestmen, daddy-long-legs)—Opiliones are an omnivorous group of arachnids that feed on detritus, plants, and fungi, with some groups constituting specialized predators. They are the only arachnids that can eat non-liquid food. Their body is characterized by fused prosoma and opisthosoma. While the familiar daddy-long-legs have very long legs, many species of Opiliones have short legs and superficially resemble mites. Some 6500 species are known, with the estimated species number being closer to 10,000.

Solifugae (sun spiders or camel spiders)—Camel spiders are relatively large and fast-moving arachnids. Most are active predators who chase their prey. They have no venom, and rely on large sharp chelicerae to catch and dissect prey animals. The pedipalps are often long sensory structures that are similar in size to the walking limbs, so that camel spiders look as though they have 10 legs. There are over 1200 species of camel spiders, ranging in size from a few millimeters to a 10 centimeter leg-span. They are notable for the racquet organs, a set of flattened projections on the underside of the fourth walking leg that are used for mechanical and chemical reception. Their large size and high running speed—often during the day—make them very startling animals, but they are in fact quite harmless to humans.

Amblypygi (whip spiders)—These are superficially spider-like animals, with flat bodies and large, well-developed pedipalps used for hunting. Only about 250 species of these relatively rare, ground-dwelling animals are known. They are named for their antenna-like "whips," which are in fact modifications of the first pair of walking legs that are equipped with an array of sensory structures. Whip spiders are functionally six-legged animals.

Evolutionary history

The earliest uncontested chelicerates are Ordovician horseshoe crabs, followed by the Silurian eurypterids, or sea scorpions. The body of both groups is clearly divided into a prosoma and opisthosoma and they have appendages that are homologous to the chelicerae and pedipalps of extant chelicerates. However, there are several older fossils from the Middle Cambrian that some workers identify as belonging to Chelicerata. The fossil *Sanctacaris* is recovered in some analyses as the earliest member of Chelicerata. However, it lacks the typical chelicerae that are the hallmarks of the taxon. A diverse group of Cambrian fossils known as Megacheira are believed to be on the branch leading to modern chelicerates, after the mandibulate–chelicerate split, although not all workers accept this interpretation.

Many fossil marine chelicerates are known from the Ordovician and onward. These mostly belong to the eurypterids, mentioned above, and to the xiphosurans. The earliest terrestrial chelicerates are the

Silurian trigonotarbids—an extinct order of small mite-like animals with four book lungs, from found in rocks in Shropshire, UK. Other arachnid orders appear in a narrow window of geological time through the Devonian, including modern-looking Opiliones from the Rhynie chert in Scotland. These fossils are believed to represent early examples of the first chelicerate terrestrialization. Until recently, this was believed to also be the only terrestrialization event within this group. However, the idea that arachnids may be paraphyletic suggests that this may be only one of two or possibly more such events. A number of fossil scorpions are known from the Silurian as well, though workers remain divided on whether these are aquatic or terrestrial fauna.

Arachnids are surprisingly well-represented in the terrestrial fossil record. There are over 1500 known fossil species, mostly spiders, preserved in amber and in other types of deposits. These fossils suggest that all modern lineages had probably diverged by the late Paleozoic (about 300–250 million years ago).

Box 21.1 Panarthropoda

The two phyla most closely related to Arthropoda are Onychophora, the velvet worms, and Tardigrada, the water bears. The three phyla together are included under Panarthropoda, making up the third major branch of Ecdysozoa (after Nematoida and Scalidophora—Chapter 20 and Box 20.1).

Members of Panarthropoda are characterized by serially repeated lateral appendages, as described above for arthropods. However, in tardigrades and onychophorans, we do not see the full body segmentation, incorporating diverse organ systems that we see in arthropods.

The limbs of tardigrades and onychophorans are also not themselves segmented. Panarthropods are characterized by an anterior brain composed of one to three dorsal ganglia, and linked to anterior sense organs. The nervous system of the trunk is a ventral cord in all three phyla. It incorporates segmental ganglia in tardigrades and in arthropods, but not in onychophorans.

Onychophorans are long and thin worms, with stubby appendages. They are all predators, hunting with specialized glands that exude a sticky substance that immobilizes their prey. They are found in forests in tropical and damp temperate regions. Onychophora is unique in being the only phylum that is composed exclusively of terrestrial species. Nonetheless, its origins, like those of all animals, are in the sea (see below).

Tardigrades are tiny animals, rarely more than a millimeter in size, that live in moist environments on land as well as in freshwater and in the sea. They have four pairs of short legs and a single-segment brain. Tardigrades are well-known for their resistance to all manner of environmental stress. They achieve this resistance by entering a dormant mode known as **cryptobiosis**, in which their bodies are nearly dehydrated and the cuticle thickens to give an almost impervious armor.

In addition to the three extant phyla, there is a diversity of fossil panarthropod taxa, which have phylogenetic positions at the base of Panarthropoda or within Panarthropoda but outside the extant taxa. Many of these are characterized by short and unsegmented appendages known as **lobopodia** and are referred to generally as "lobopods"—although this is likely a paraphyletic assemblage. Some of these look very much like marine onychophorans, and may be their direct ancestors. Others look like elongated versions of tardigrades and indicate that the short and compact body of tardigrades was achieved through miniaturization of an ancestral lobopod body plan. Fossil panarthropods include some of the "weird wonders" of the Cambrian, including *Hallucigenia*, *Opabinia*, and *Anomalocaris* (see Box 12.1), as well as many other species. Studying these fossils has led to much better understanding of the early evolution of arthropods and their close relatives.

CHAPTER 22

Terrestrialization

Life began in the sea

The sea is the cradle of life on Earth. The first living organisms appeared in the sea some 3.5 billion years ago. Multicellular life also evolved in the sea, as did the first bilaterians. The origins of all extant phyla are in the sea and all phyla have marine representatives, with the exception of Onychophora. It is therefore no surprise that the physiology of all extant animals was originally adapted to the marine realm. The composition of minerals in the body fluids of animals closely matches the composition of sea water. Systems for acquiring oxygen are adapted to absorbing oxygen dissolved in water. The first muscles and skeletal systems evolved for movement within fluid surroundings. Leaving the sea thus required significant modification and adaptation of almost all organ systems. Only a handful of phyla made the transition to land, and only three or four of these have fully adapted to terrestrial environments.

Constraints to living in the air

While we tend to think of **terrestrialization**—the adaptation to terrestrial habitats—as a move from sea to land, the challenging aspect of terrestrialization is not land but air. Living in an environment surrounded by air imposes a suite of challenges that the organism must adapt to.

The most obvious difficulty in living in the air is the constant threat of water loss and desiccation. All animal tissues are composed largely of water, including cytoplasm, intracellular liquids, and fluid-filled cavities, such as coeloms and hemocoels. Losing water to the environment disrupts the function of all these water-based fluids, mostly by increasing the concentration of dissolved materials and disrupting the correct composition of minerals and organic molecules. Transitioning to air requires evolving a mechanism to conserve water and prevent evaporation. This is best done by evolving an impermeable integument that covers all or most of the body, and by evolving physiological mechanisms for maintaining a correct balance of dissolved material even in the face of water loss.

A less obvious, but no less important, constraint to living in the air is the loss of buoyancy. Most aquatic animals have a specific gravity that is close to that of water or only slightly higher. This allows most of their weight to be supported by buoyancy. In the air, all of the animal's weight must be supported by the skeletal and muscular system. Being able to move on land required evolving a locomotory system that can deal with the animal's full weight. Water viscosity is also significantly higher than the viscosity of air, and many aquatic animals move using the resistance of water as a counter force. All animals that swim or move using flippers, paddles, or fins push against the water. This is not possible in air.

Surprisingly, breathing is actually more difficult in the air than in water. Oxygen transport systems have evolved to deal with dissolved oxygen. Breathing in the air requires dissolving the oxygen in water first. Since having large breathing surfaces covered in water would be very costly in terms of preventing water loss, most terrestrial animals evolved damp internal cavities that are connected to the outside air through a relatively narrow opening and where gas exchange takes place. These cavities are

Organismic Animal Biology. Ariel D. Chipman, Oxford University Press. © Ariel D. Chipman (2024). DOI: 10.1093/oso/9780192893581.003.0022

usually referred to as **lungs**. Nonetheless, smaller animals with lower oxygen requirements can take advantage of the higher concentration of oxygen in the atmosphere relative to the sea and survive by absorbing oxygen directly through epithelial cells. In these cases, the epithelial cells are buried deep within spaces connected to the outside world via small opening that can be closed when water loss is high. These opening are usually known as **spiracles**.

Terrestrial environments have much less stable temperatures due to the lower heat capacity of air. Water temperature usually fluctuates very little over a 24-hour period, as compared to air temperature that can change dramatically between daytime and nighttime. Moving to terrestrial habitats requires adaptation to this thermal fluctuation. This adaptation can be through evolving better insulation in the integument, through behavioral adaptations (e.g., returning to water during the hottest time of day) or by having a flexible physiology that can cope with differences in body temperature.

A common mode of reproduction in many aquatic animals is external fertilization via the simultaneous release of ova and sperm into the water column. This type of reproduction is not possible in air. Therefore, animals that undergo terrestrialization must first evolve internal fertilization or continue to return to water for breeding even after terrestrialization.

It is worth pointing out that even a transition from sea water to fresh water carries physiological constraints, which partially mirror the constraints of transition to air. Fresh water has a lower concentration of soluble minerals, leading to the opposite danger to that of transition to air: a decrease in the concentration of dissolved minerals and organic molecules due to influx of water from the environment. Fresh water bodies tend to be smaller and as such have less stable temperatures than the sea (though not as variable as air). Finally, fresh water tends to have lower dissolved oxygen concentration than sea water. These constraints may seem irrelevant to the question of terrestrialization, but as we will see in the next section, adaptation to fresh water can be a first step toward the transition to land. Plants have also adapted to the challenges of living on the land as described in Box 22.1.

The routes to terrestriality

Studies in comparative and evolutionary physiology suggest that there are two common routes to terrestrialization. Some taxa started by adaptation to littoral areas (areas near the shore) and tidal zones where salinity levels fluctuate over time. Others first adapted to estuarine areas where fresh water meets the sea and from there transitioned to fresh water and ultimately to land.

Before discussing this further, we need to define two ways of dealing with varying salinity levels (see Chapter 28 for more details). Organisms are said to be **osmoconformers** if they match their osmotic balance (the concentration of soluble materials in their body fluids) to that of the environment. Osmoconformers that can tolerate fluctuations in the surrounding water are better adapted to the unstable conditions of the littoral and tidal environments. Organisms that are **osmoregulators** are able to maintain a constant osmotic balance regardless of the osmotic level of the water they are in. Osmoregulators can cope with unstable conditions as well as osmoconformers: to do this requires energy, and these animals therefore have to invest more resources into maintaining their internal environment. However, they are much better adapted to very low mineral concentrations such as those found in fresh water, which would be lethal to osmoconformers.

The physiology of terrestrial animals maintains hallmarks of osmoregulation or osmoconformism, providing a hint of the evolutionary history of their terrestrialization process. Combining comparative physiology and phylogenetic analyses suggests that arthropod terrestrialization was achieved via a littoral route (with the possible exception of insects), whereas flatworms and earthworms evolved via a freshwater route, and continue to rely on the presence of fresh water for their survival. Vertebrates left the sea via fresh water, and on the way, evolved an air breathing system to cope with decreased oxygen levels, before fully leaving the water.

Full terrestrialization is rare

Of the animals that have undergone the transition from sea to land, only members of three phyla have adapted fully to living on land: vertebrate chordates (Chapters 29–31), gastropod mollusks (Chapter 15), and several lineages of arthropods (Chapters 21 and 23). These animals carry out their entire life cycle on land, including laying eggs in terrestrial environments, and are not reliant on external water to prevent desiccation. This full adaptation requires a water-resistant integument, terrestrially adapted eggs, and an excretory system adapted to conserving water.

It is interesting to note that two of these taxa, arthropods and vertebrates, evolved supportive muscle and skeletal systems and walking limbs before leaving the water. This suggests that perhaps the locomotory constraints are significant limiting factors to terrestrialization. Arthropods evolved protective cuticles early in their history, providing an existing adaptation for prevention of water loss. Vertebrates evolved a water-resistant integument only after undergoing terrestrialization, but their large size may have provided a buffer against rapid desiccation.

Mollusks are surprising as terrestrial animals, given their soft and permeable integument. There is some debate over the question of how many times gastropods underwent terrestrialization and what route (or routes) they took. The mantle cavity, surrounded by the water-resistant shell, provides a hiding place from desiccation during hot and dry periods. The air-breathing organs of pulmonate gastropods probably evolved in fresh water, providing an additional advantage when venturing out to land. Interestingly, some terrestrial gastropods, the slugs, lost the shell despite its apparent advantages. However, slugs are confined to damp and humid environments and rarely venture out during the heat of the day. Those slugs that live in warmer climates go into summer dormancy or **estivation** to avoid desiccation.

Partial terrestrialization

Members of several other phyla are found in terrestrial environments, but they, like the slugs, are confined to environments with high humidity or proximity to water. Such environments include rain forests, swamps, damp leaf litter, or water-logged earth. The eggs of these animals are usually not desiccation resistant and must be laid in or near water. Examples of these not fully terrestrial taxa include flatworms, earthworms, onychophorans, tardigrades, and nematodes. In many of these taxa, there have been several terrestrialization events, including secondary reversals to aquatic environments. In almost all these cases, the route to land went through fresh water.

Evolutionary history of land animals

The first terrestrial animals were undoubted arthropods. Molecular clock evidence suggests that myriapods (centipedes and millipedes) and possibly also arachnids (spiders and their kin) ventured on to land as early as the late Cambrian or early Ordovician (roughly 500 million years ago). This is corroborated by the earliest terrestrial arthropod trackways, probably created by myriapod ancestors at about the same time. Arachnids are believed to have undergone terrestrialization in a single event, but there is some evidence to suggest this may have happened more than once (see Chapter 21). The evolutionary history of insect terrestrialization is unclear, but it must predate the earliest insect fossils from the Devonian (385 million years ago). Several lineages of crustaceans underwent terrestrialization separately, most of these were much later than the terrestrialization of other arthropods. Fossil terrestrial isopods (woodlice) are known from the Cretaceous (roughly 100 million years ago). Land crabs may have left the sea only in the last few million years.

The evidence for vertebrate terrestrialization is much clearer (see Chapter 30). Most adaptations to life on land took place in organisms that were still mostly aquatic. A sequence of beautiful fossils spanning the late Devonian illustrate the gradual refinement of walking limbs and other locomotory adaptations from air-breathing fish-like animals to amphibious tetrapods (four limbed vertebrates) that spent most of their life on land.

The evolutionary history of other terrestrial taxa (gastropods, flatworms, etc.) is much less clear, since they have almost no relevant fossil record and there isn't even a consensus about the number of terrestrialization events in the different lineages, and even less so about the timing of these events.

Box 22.1 Plant terrestrialization

Although this book focuses on animals, we cannot ignore the importance of plant evolution to the evolution of animals. The major events in plant evolution lagged behind those of animals for millions of years. However, once plants invaded land, they underwent a rapid increase in diversity and complexity and became established as a significant element of the biosphere.

The ancestors of today's land plants lived in freshwater bodies, probably along the margins of land. It is in these bodies that they evolved complex multicellularity, including the earliest leaf-like structures and roots. The first plants left the water in the late Silurian. Unlike the multiple terrestrialization events in animals, the evidence points toward plant terrestrialization being a singular event. In other words, all terrestrial plants found today are descended from a single lineage of Late Silurian terrestrial plants.

The establishment of plants on land provided new resources and niches for the recently terrestrialized animals and probably facilitated their establishment and diversification. It is important to point out that there were other photosynthetic organisms that made the transition to land even earlier, though these were probably not multicellular and their impact on the terrestrial environment was minor. Nonetheless, they probably provided nourishment for the earliest terrestrial animals.

Plants of the early Devonian, a few million years after their first terrestrialization, were still small, had rudimentary root systems, and almost no transport systems. These early plants were structurally similar to the mosses of today. However, throughout the Devonian, plants underwent dramatic evolutionary changes, leading to increased diversity and complexity. By the end of the Devonian, the world was covered in forests of tall, woody fern-like plants. These forests transformed the face of the planet and provided the setting for the evolution of terrestrial animals.

CHAPTER 23

Arthropoda II

Mandibulata

Mandibulate body organization

Mandibulata is the larger of the two branches of arthropods. It includes three Linnaean classes: myriapods (centipedes, millipedes, and their relatives), hexapods (insects and their relatives), and crustaceans (crabs, shrimp, lobsters, and their kin) (see Figure 23.1). Molecular phylogenetics have made it clear that crustaceans are a paraphyletic group, with hexapods as just one of the branches within them. Therefore, the taxonomic name Pancrustacea is used to refer to the clade including hexapods and crustaceans. Mandibulates are mostly clearly characterized by the presence of a distinct head (also known as the **cephalon** [see Box 23.1]), which is devoted to feeding and to sensory functions, with no locomotory appendage beyond the larval stage (unlike the prosoma of chelicerates, which has locomotory appendages in addition to feeding and sensory structures). The cephalon is not always clearly delimited from the rest of the body in crustaceans, but in hexapods and myriapods it is usually enclosed in a distinct head capsule. The head is followed by a series of trunk segments, which may be differentiated into two or more distinct tagmata. At least some of the trunk segments bear locomotory appendages, specialized for walking or swimming, whereas others can bear other specialized appendages or have no appendages (see Figure 21.1).

The mandibulate head includes three paired ventral feeding appendages, each pair carried on a separate segment. The anteriormost paired appendages are the eponymous **mandibles**. These are usually heavily sclerotized and function in cutting or crushing food items. The second pair is the **maxillae**, which usually function as accessory feeding structures. The third pair is sometimes fused, and the fused structure is then known as the **labium**. If the pair is not fused, they are known as the second maxillae. The feeding appendage bearing segments are known as the **gnathal segments** and the appendages they bear are thus sometimes referred to as gnathal appendages. The exact structure of these three feeding appendages is highly variable, especially in insects, and is specifically adapted to the organism's food source and mode of feeding. Some of these may be reduced or fused, to give a specialized complex feeding apparatus.

Anterior to the segments bearing feeding appendages are three segments that bear sensory structures. The anterior most segment is the **ocular** segment which bears the eyes. In mandibulates, these are usually compound eyes, which may or may not be carried on a stalk. The anterior segment also bears an appendage-like structure, known as the **labrum**, which functions as a protective cover over the mouth and feeding appendages. The second segment is the antennal segment, and this bears thin sensory appendages known as **antennae**. The antennae are made up of repeated **articles** that are usually similar in shape. The number of articles is highly variable, as is the degree of differentiation among the articles. Crustaceans have two pairs of antennae, and the anterior ones (those on the second segment) are known as **antennules**. The third segment bears the second pair of antennae in crustaceans. Hexapods and myriapods have secondarily and convergently lost the second antennae, and the limbless third segment in these taxa is

Organismic Animal Biology. Ariel D. Chipman, Oxford University Press. © Ariel D. Chipman (2024). DOI: 10.1093/oso/9780192893581.003.0023

known as the **intercalary** segment. The mandibulate head is thus composed of six segments: three pre-gnathal segments and three gnathal segments.

Evolutionary history

The Cambrian fossil record includes species that already have unquestionable crustacean affinities. Most notable, a series of superbly preserved fossils from the late Cambrian Orsten deposits in Sweden includes larval stages of crustaceans, some of which can be assigned to extant orders. As is the case with chelicerates (see Chapter 21), there are some earlier fossils that some workers believe to be on the lineage leading to modern mandibulates, after the chelicerate–mandibulate split, but before the divergence of the extant classes. The most famous of these is *Fuxianhuia* from the Chengjiang deposits in south China. *Fuxianhuia* is also notable for its excellent preservation, with some specimens preserving neural tissue, circulatory organs, and digestive tracts, giving us an unparalleled view into early arthropod anatomy.

Terrestrial myriapod fossils, both body fossils and putative trackways, are known from roughly the same time as the earliest terrestrial arachnid fossils late in the Silurian period. The fossil euthycarcinoids are the closest marine relatives of the myriapods. The closest marine sister group to hexapods is unknown in the fossil record. Molecular phylogenies identify remipedes (see below) as their closest relatives, but unfortunately there is no fossil record for this taxon.

Insects have an excellent fossil record, much of it in amber. This record allows us to trace the evolution of most modern insect taxa and of their extinct close relatives. However, the early evolution of insects, reflecting their emergence from within Hexapoda, and the emergence of Hexapoda from Pancrustacea is almost entirely unrepresented in the fossil record.

Myriapod diversity

Myriapoda is the least diverse of the four arthropod classes, both in terms of species number and in terms of morphological diversity. Myriapods have a relatively simple body plan, with only two tagmata:

the head and the trunk. The trunk is composed of numerous leg-bearing segments that are almost all identical. This simple body plan is not necessarily primitive and may represent a secondarily simplification from an ancestor with more diverse segment morphologies. Myriapods are entirely terrestrial, with no known aquatic species. They breathe through a pair or a series of spiracles connecting to a network of tracheae. The tracheae end internally in thin walled, liquid-filled regions, where gas exchange takes place directly with the hemolymph. Myriapods are important components of the soil fauna in almost all biomes, their long and thin bodies allowing them to burrow into the ground, inhabit cracks in the soil, or hide under rocks.

There are four orders included with in Myriapoda. Two of these are fairly significant, Diplopoda and Chilopoda, and are discussed in somewhat more detail below. The two others, Symphyla and Pauropoda, include relatively few species of small soil-dwelling animals that are poorly known and not frequently seen (unless one looks for them specifically), and we will not discuss them further here.

Chilopoda (centipedes)—Active predators, mostly specialized for hunting various soil invertebrates. There are about 3000 described species. Centipedes are characterized by limbs that point laterally for running. This is most extreme in the Scutigeromorpha, the group that includes the familiar house centipedes, with their very long sideways-facing legs. The antennae are usually long and function as mechanosensory and chemosensory appendages. The first trunk segment of the centipedes is modified to bear a pair of venom claws or **forcipules** (sometimes also **maxillipeds**, but note that this term is also used for a non-homologous structure in crustaceans). The venom claws are sclerotized curved appendages with a sharp point that are connected via a venom duct to unique glands situated on the first trunk segment, behind the head. Centipedes hunt by catching their prey with the venom claws and injecting them with a toxin that immobilizes or kills them. The rear limbs are also modified and point posteriorly, functioning as additional mechanosensory appendages, and sometimes as defensive structures.

All adult centipedes have an odd number of leg-bearing segments (not counting the venom claws). Some orders, Lithobiomorpha and Scutigeromorpha, have a constant number of 15 leg-bearing segments (30 legs). Members of Scolopendromorpha nearly always have 19, 21, or 23 pairs. Members of Geophilomorpha can have anything from 27 to nearly 200 pairs of leg-bearing segments, with the number varying even within a species. It is worth pointing out that although the name "centipede" literally means 100 legs, there are no centipedes with 100 legs, since that would require 50 leg-bearing segments—an even number. Those centipedes with 15 leg-bearing segments are anamorphic (adding segments post-hatching) and those with higher numbers are epimorphic (generating all their segments during embryogenesis). Some species of centipedes are active on the surface, running and hunting their prey in the open (most notably scolopendromorphs), some live primarily in leaf litter (e.g., lithobiomorphs), where they can be top predators of a diverse arthropod community, whereas others, such as the worm-like geophilomorphs, are subterranean, emerging only rarely.

Diplopoda (millipedes)—Slow moving detritivores and herbivores with more than 13,000 described species. Millipedes tend to have more legs than centipedes (although there is an overlap). Unlike centipedes, millipedes' legs point ventrally. Most segments are arranged into **diplosegments**, each being a fusion of two adjacent segments and bearing two pairs of legs. The name Diplopoda ("double legs") comes from this unusual arrangement. Millipedes do not have poison claws. The antennae are generally shorter than those of centipedes, and they do not display the posterior-facing sensory legs that centipedes have. All millipedes have anamorphic development. The shortest millipedes have 14 pairs of legs at the adult stage, while the longest millipedes have many hundreds of legs. There is only one known example of a millipede with more than 1000 legs ("millipede" translates as 1000 legs), an Australian species with up to with 1300 legs.

Most millipedes have highly sclerotized or calcified cuticles. In many taxa, this cuticle forms a cylindrical external skeleton, made up of a series of rings, giving them a worm or tube-like appearance. Most long millipedes curl up to a disk-shape for defense, whereas other shorter species can roll up to give an almost spherical shape (so-called pill millipedes). The millipede cuticle is lined with repulsive glands, which open to **ozopores**. These glands excrete nasty volatile substances that deter predators, adding an additional layer of defense.

Crustacean diversity

Crustaceans are a hugely diverse group of arthropods. As mentioned above, they are technically a paraphyletic group, with the correct taxonomy including hexapods as a group with the monophyletic Pancrustacea. However, most traditional taxonomies ignore this, and we will follow the traditional approach and discuss them separately. When crustaceans are ranked as a subphylum, they are traditionally divided into 10 or more classes and close to 50 orders, a testament to the morphological diversity within this group. All crustacean classes are primitively aquatic, but partial or complete terrestriality has evolved several times within different classes. Even those crustaceans that have moved to life on land still bear the hallmarks of their aquatic ancestors, most notable, breathing through modified gills and not tracheae. No crustaceans have achieved full terrestriality, with the exception of some members of Isopoda. Many crustaceans have biphasic life histories with typical nauplius larvae, and occasionally a varying number of additional taxon-specific intermediate larval and preadult forms.

The crustacean body plan is composed of a typical mandibulate head, with two pairs of antennae. The head is often fused to the tagma posterior to it via a carapace that encompasses both the head and several trunk segments. This fused structure is known as a **cephalothorax**. The region that includes segments that are free from the carapace is known as the **abdomen**. When the head is not fused to posterior trunk segments, the tagma that includes locomotory appendages is known as the **thorax**. The identity of the limbs of specific segments is highly variable and is often typical for a given order. Limbs can be specialized for walking or swimming, for handling food (when they are known as **maxillipeds**), for carrying eggs and for other activities. In some crustaceans the appendages include more highly or less developed gill elements, whereas in others there are dedicated appendages that function solely as gills.

The diversity of crustacean taxa and the variations on their body plan are too complex to be discussed fully here. We therefore list a sample of only the best-known and most important crustacean orders.

Malacostraca—This large taxon, with over 40,000 species, includes the most familiar crustaceans. Among others it includes crabs, lobsters, prawns, shrimp, krill, and similar animals. Note that the term "shrimp" appears in the colloquial names of many crustaceans that are not true shrimp and many are not even malacostracans (e.g. fairy shrimp, brine shrimp). Malacostraca includes Isopoda, the woodlice and sea slaters, which has both terrestrial and marine representatives. Terrestrial isopods are the crustaceans with the most extreme adaptations to life on land, with some species even adapted to life in the desert. Decapoda is the taxon within Malacostraca that includes lobsters, crabs, and other ten-legged crustaceans. There have been several convergent events of "carcinization" within Decapoda. Carcinization is the evolution of a crab-like body plan—a flat and rounded body with the abdomen folded ventrally—usually associated with life on the sea floor. Malacostracans usually have a carapace covering the head and at least some of the thorax. The thorax or **pereon** is composed of eight appendage-bearing segments. In decapods, the anterior three are modified as maxillipeds, leaving five pairs of locomotory appendages. The abdomen or **pleon** is usually composed of six segments that have non-locomotory appendages. The pleon is thick and muscular in many species, but in carcinized species it is very small and folded under the thorax. The posterior-most structure is known as the **telson** and in non-carcinized species it often bears a tail-fan.

Cirripedia (barnacles)—The barnacles are highly-specialized crustaceans that adopt a sessile lifestyle in the adult stage. Indeed, they are so highly modified relative to the ancestral crustacean body plan that they were only recognized as crustaceans in the mid nineteenth century, previously being allied with mollusks. There are about 1200 described species, including the familiar acorn barnacles, with a tower-shaped shell-like exoskeleton covering the entire body and hiding most of the animal's morphology, and the goose barnacles that are connected to the substrate with a thick pedestal. Cirripedia also includes the parasitic barnacles, bizarre animals that live inside crabs and other malacostracans, and have lost nearly all trace of arthropod morphology, being more like a cancerous branching tissue growing inside their host.

Branchiopoda—This is another familiar taxon with over 1000 described species. It includes small planktonic crustaceans such as fairy shrimp, (including brine shrimp or sea monkeys), tadpole shrimp, and water fleas. Their name means gill-legs, and indeed their appendages include significant respiratory structures. Branchiopods swim—often ventral-side up—using their flattened appendages as oars and as feeding structures for filtering food out of the water. In clam shrimp, the whole body is covered by a bivalved carapace. Branchiopods are found mostly in fresh and brackish water, but some taxa are also found in the sea. They are often numerous and are an important component of the base of the food chain in these environments.

Copepoda—Another group of small planktonic crustaceans, found both in the sea and in fresh water. Copepods are a major component of plankton, with over 14,000 described species. Some estimates claim that they are the most significant component of marine plankton, both in number of individuals and in total biomass, although such estimates are difficult to confirm. Most copepods swim using their small thoracic appendages as oars, their name literally meaning oar-feet. Copepods have a cephalic shield but no extended carapace. They have a reduced thorax with only six segments and no appendages. About half of copepod species belong to lineages that have evolved a parasitic lifestyle.

Ostracoda (seed shrimp)—These small crustaceans have a simplified body plan, with reduced segments and weakly defined borders between the segments. Their body is covered by a bivalved carapace, giving the appearance of tiny swimming clams. Although they share a bivalved carapace with the clam shrimp, the two are not related, the

(a)

(b)

(c)

(d)

Figure 23.1 Examples of mandibulate diversity: (a) A centipede, *Scolopendra cingulata*. (b) An isopod crustacean *Porcellio buddenlundi*. (c) A decapod crustacean, the ghost crab *Ocypode quadrata*. (d) A hemimetabolous insect, the praying mantis *Mantis religiosa*.

Source: a: Photo supplied from Shutterstock: Erni (https://www.shutterstock.com/image-photo/scolopendra-cingulata-known-megarian-banded-centipede-1393310828); b: Photo supplied from Shutterstock: skippy666 (https://www.shutterstock.com/image-photo/porcellio-buddelundi-woodlice-isopod-assel-soil-1625829067); c: Photo supplied from Shutterstock: MarynaG (https://www.shutterstock.com/image-photo/ghostcrab-ocypode-quadrate-sunlight-on-491429005); d: Photo supplied from Shutterstock: Tatyana Sanina (https://www.shutterstock.com/image-photo/female-european-mantis-praying-religiosa-green-1331151110).

former belonging to Branchiopoda. There are about 7500 identified extant species, but more than 50,000 fossil species. Because their carapace fossilizes readily, ostracods are among the most common arthropod fossils, and are often used for sedimentological analyses and for biostratigraphy (the alignment of different rock strata based on their biological components).

Remipedia—Only about 25 species are known from this enigmatic taxon, and all are found in deep marine caves. They are unusual in having a homonomous trunk, meaning all segments have the same general morphology and bear similar appendages. Remipedia is worth mentioning

because most phylogenies recover it as the sister group to Hexapoda. Because remipede morphology and ecology are so unusual, and so different from the morphology of their sister taxon, it is likely that they represent a secondarily simplified body plan.

Pentastomida (tongue worms)—highly derived parasitic crustaceans. They were once thought to be a separate phylum, until sperm structure and then molecular phylogenies identified them as crustaceans. They have worm-like bodies and are usually parasites of the respiratory system of terrestrial vertebrates, where they attach using their head appendages (the only appendages in the body) and feed on the host's blood or mucus.

Hexapod diversity

With over a million described species, hexapods are among the most common and best-known animals on the planet. The vast majority of hexapods belong to Insecta, with the non-insect hexapods comprising only 9000 species in three orders (of course, "only" is relative to insects). Despite this huge diversity in species number, the hexapod body plan is conservative. It is composed of three tagmata: a head, the typical mandibulate head described at the beginning of this chapter; a thorax, comprised of three segments, each bearing a pair of walking appendages (giving the hexapods their name, literally six-legs); and an abdomen composed of 8–11 segments mostly with no appendages. Most insects have two pairs of wings, situated dorsally on the second and third thoracic segments. Unlike most terrestrial arthropods, the hemolymph does not have a role in gas exchange, and branches from the tracheae reach every cell in the body to deliver oxygen and remove carbon dioxide.

Hexapods are primitively terrestrial, almost definitely representing a single terrestrialization event early in their history. Some insects have secondarily aquatic larvae and rarely, aquatic adults. Insects are found everywhere on land and in fresh water, often in large numbers. They are major components of any ecosystem in which they are found. There are predatory hexapods, herbivorous hexapods, detritivores, parasites, and numerous variations within these feeding types. Indeed, the diversity and importance of insects is such that their study is defined as a distinct discipline, entomology, with its own academic departments, journals, and scientific conferences.

The ecological diversity of hexapods is manifested mostly in three structures: The walking appendages, the feeding appendages, and (in most insects) the wings. Walking limbs can be specialized for walking, running, climbing, jumping, digging, prey capture, and other tasks, each specialization affecting the size and shape of the legs and the relative proportions of the segments that compose them. Feeding appendages can be specialized for chewing, biting, piercing, sucking, licking, and other ways of feeding, with the three pairs of gnathal appendages adopting myriad shapes and

relative positions to achieve these specializations. Finally, the wings can be adapted for gliding, slow flight, rapid maneuverable flight, hovering, and, in some cases, for protection.

Once again, the diversity of hexapods is far beyond what can be accommodated in this section, so the discussion necessarily leaves out many important aspects. We will now discuss a few representative orders, pointing out how they are adapted for specific ecological niches.

Non-insect hexapods—Sometimes referred to as Entognatha, although this is a paraphyletic assemblage. The non-insect hexapods lack some of the diagnostic characters that are found in insects (e.g., flagellar antennae with long muscle strands), and the conditions seen in Entognatha presumably represent the primitive state. The character that gives this assemblage its name is the fact that the base of the feeding appendages is hidden within cheek-like pouches in the head capsule. The main order of non-insect hexapods is Collembola, the springtails, with about 8100 described species. These are small animals, usually a few millimeters in length, which are found in the soil, in leaf litter, or on the surface of standing fresh water. They have a special ventral appendage on the abdomen, known as a **furcula**, which allows them to jump to many times their body length. Unlike all other hexapods, most of them breathe directly through the cuticle and do not have tracheae.

The other two non-insect orders are Protura and Diplura, each with a few hundred species. Both are comprised of small soil animals. Protura have no eyes and no antennae, and have the distinction of being the only hexapods with anamorphic development.

Archaeognatha (bristletails) and Zygentoma (silverfish and firebrats)—These are the primitively wingless insects. They are sometimes grouped together in the probably paraphyletic Thysanura[*]. Approximately 450 species are known. They are small- to medium-sized insects with a flattened body. They are characterized by three rear-facing mechanosensory appendages known as **cerci**, which are mostly used to detect air movement. Unlike other insects, their body is covered in delicate scales.

The winged insects or Pterygota are divided into hemimetabolous insects and holometabolous insects, which are differentiated by their life history. Hemimetabolous insects hatch as **nymphs** which have a generally similar body plan to that of the adults. They grow through successive molts, with each instar being more similar to the adult. In the final molt they develop wings (in most cases) and external reproductive organs and become sexually mature. There are no molts in the adult stage.

Holometabolous insects hatch as larvae that are very different from adults. They often have no legs and if they do, these are rudimentary or very different from adult legs. The head and feeding appendages are usually different, representing different feeding ecologies between the larvae and adults. After several larval instars the larvae turn into **pupae,** which form a non-feeding, usually immobile life history stage, during which the animal undergoes a dramatic metamorphosis, with many structures being broken down and replaced by new structures. The adult or **imago** that hatches from the pupa has wings and reproductive organs and does not molt further. **Hemimetabolous** insects are a paraphyletic group representing the primitive life history of insects. Holometabola is a monophyletic group, and metamorphosis is a novel character within this group. The holometabolous life cycle apparently provides some significant advantage, since the majority of insect species are found within this group.

Some of the more important hemimetabolous orders (out of the 15 or so recognized orders) are detailed below.

Ephemeroptera (mayflies)—The 2500 species of mayflies display many primitive characters in the structure of their wings and of their mouthparts. The larvae are aquatic, usually living in fresh running water, and are the dominant life history stage. The adults live for only a few days, and their ephemeral adult existence gives the order its name. Mayflies are unique among insects in having a winged pre-adult stage. The pre-adult (or sub-imago) emerges from the last larval molt, settles on a branch or other protrusion, and after a few hours molts again to reach the final adult stage.

Odonata (dragonflies and damselflies)—Like mayflies, the dragonflies and damselflies have aquatic larvae, but the adults live for longer and are a more significant stage. They also share some primitive characters in the wing joints and are thus joined with Ephemeroptera under a monophyletic group known as Paleoptera (ancient wings). Both groups are predatory both in the larval and in the adult phase. Larval odonates are often top predators in ponds and temporary pools, feeding on other insects, frog tadpoles, and other prey. As adults, they are airborne hunters, with large compound eyes allowing them to chase and catch flying insects. There are approximately 6000 known species of these graceful insects.

Blattodea (cockroaches and termites)—Although many of us are most familiar with the human associated pest species, cockroaches are actually a diverse group with representatives in many different habitats. They have wings that fold over the body to provide protection and are usually not useful for active flight. Indeed, many cockroach species have lost their wings. They have chewing mouthparts and are usually generalist feeders. Termites are a group within cockroaches that have evolved a social lifestyle, living in nests with only one or a few reproductive individuals and thousands to millions of non-reproductive workers and soldiers. The reproductive individuals shed their wings after mating, and the non-reproductive ones never grow them. Many termites feed on wood and host endosymbiotic dinoflagellates (see Chapter 3) to help them digest the complex molecules of which it is composed. There are about 4000 species of cockroaches and 3000 species of termites. In terms of biomass, termites make up a major component of several ecosystems, include savannahs and steppes.

Mantodea (mantises)—All of the 2400 species of mantises are predators, and their anterior legs are modified as hunting appendages rather than walking legs. As active predators, they have a flexible joint between the head and thorax, and large forward-facing eyes, allowing them to track prey. Their mouthparts are dominated by large powerful mandibles. Mantises are generally poor fliers and many species have lost their wings. Cockroaches and mantises share a similar way to protect their offspring, in which eggs are deposited into an egg case, which is either carried by the female or left in a suitable location. After embryogenesis, the first instar nymphs hatch out of the egg case.

Orthoptera (grasshoppers and crickets)—a diverse order with some 23,000 described species. Orthopterans are characterized by enlarged posterior legs, usually used for jumping. The forewings are mostly used as a cover for the hindwings, which can be large and membranous. In some cases (crickets) the wings are used to produce sound. Some orthopterans are excellent fliers, for example, the locusts, which can fly for hours at a time covering great distances. Other orthopterans use their wings only to assist and extend their jumps. There are also species that have lost their wings entirely. Orthopterans are nearly all herbivorous or omnivorous, but there are also a few carnivorous species.

Hemiptera (bugs)—While the term "bug" in colloquial usage can refer to almost any arthropod, in zoological usage it is reserved for this group of insects, characterized by mouthparts that are modified for piercing and sucking. Hemiptera is the largest hemimetabolous order with some 85,000 described species. All hemipterans live on a liquid diet. This can be plant liquids (from phloem and xylem) in the case of aphids, cicadas, shield bugs, and other plant bugs. Other bugs feed on blood (the much-reviled bedbugs) or insect hemolymph (assassin bugs). Seed bugs use their mouthparts to inject digestive enzymes into seeds and then suck up the liquified product. Hemipteran forewings are often reduced or used as a cover for the hindwings. Only the anterior half of the wing is sclerotized, hence the name, meaning half-wings. The legs are usually generalized walking legs, although some species have modified hind legs specialized for jumping.

There are about 10 recognized holometabolous orders. We will only list the "big four," each with over 100,000 species, which between them account for about half of the world's recognized biodiversity.

Hymenoptera (ants, wasps and bees)—Many of the 120,000 or so species of this group have advanced sociality. This is a fascinating biological phenomenon wherein the colony functions as a "super-organism," with division of labor and only one or a few reproductive individuals. This is especially interesting, since it has evolved numerous times within this order (as well as in termites, discussed above). Intermediate levels of sociality and communal nests are found within some wasps and bees. Ant nests in the tropical forest can reach millions of individuals, with colonies sometimes merging to give super-colonies with hundreds of millions of individuals. A honey bee hive can have tens of thousands of workers. Hymenopterans are small- to medium-sized insects. With the exception of ants, where most individuals are wingless, they are excellent fliers, with short but highly maneuverable wings. Feeding modes are variable and include predatory species (many wasps), pollen and nectar feeders (bees), and generalists (most ants). The structure of the mouthparts is equally variable, matching the main food source. Many wasps are parasitoids, laying their eggs inside or on other species—usually arthropods—where the embryos hatch and the larvae consume the host from within. Parasitoid wasps may represent the largest diversity within Hymenoptera, and it is believed that this diversity is only beginning to be revealed.

Coleoptera (beetles)—Roughly one out of every three species on Earth is a beetle. With 380,000 described species and an unknown number of undescribed species, the diversity of beetles is unparalleled. As one can guess, this diversity is matched by a range of feeding habits and specializations. There are carnivorous beetles, herbivores, seed specialists, dung eaters, detritivores, pollen eaters and many others. Nonetheless, beetles have a number of characters that unite them and indicate that they are a single lineage. Most beetles tend to have fairly generalized forward or ventral-facing mouthparts. The thick and sclerotized forewings are known as the **elytra**. They are sometimes fused, forming a tough shield, and preventing flight, but more often they are movable, revealing the membranous hindwings.

Diptera (flies and mosquitoes)—This order's name (two-wings) comes from the fact that the hindwings are reduced and function as balance organs (the **halteres**) and only one pair of wings is used for flight. Dipterans are very significant for human health and wellbeing, with mosquitoes being responsible for transmitting many of the world's deadliest diseases, and flies contributing to poor sanitary conditions. Dipterans are highly maneuverable fliers, with large compound eyes providing excellent vision. The 135,000 species present a diversity of feeding habits and corresponding mouthpart morphology. There are carnivores, blood suckers, detritivores, scavengers, and more within this group.

Lepidoptera (butterflies and moths)—Only a small proportion of the 120,000 lepidopteran species are colorful, day-flying butterflies. Most of the diversity of this group lies within indistinct night-flying moths. As a rule, adult lepidopterans are nectar feeders, with tube-like sucking mouthparts. Most of the lifespan is spent in the larval phase, with the larvae (commonly known as caterpillars) being ecologically more diverse and feeding on a range of plants, often with strong preference for one or a few species. The wings are large and covered with scales, which gives the order its name (scale-wings). Many butterflies are slow fliers, but some moths can cover long distances in a single night's flight. Some butterflies—most famously the North American monarch butterflies—migrate for long distances.

Box 23.1 The arthropod head

The evolution of the arthropod head has been called "the endless debate." Indeed, the structure of the arthropod head and the homology of anterior structures in different arthropod lineages have been debated extensively for over a century. The reason for the debate is clear when one looks at the anterior of the body in the four classes.

Insects and myriapods have a distinct head capsule, and there is almost no question as to what the head is. In crustaceans, the head is normally fused to the trunk and there is no clear head capsule. Chelicerates do not have a head at all, and the functions of the head (feeding and sensation) are performed as part of the prosoma, which also functions in locomotion.

As endless as the debate may have seemed a few decades ago, there is now an emerging consensus regarding the homology of the anterior structures of the arthropod body. This consensus is supported by a range of data sources, including embryology, gene expression patterns, neuro-anatomy, and the fossil record. The accepted homology of arthropod anterior segments is summarized graphically in Figure 21.1.

The consensus sees the three anterior-most segments as homologous across all arthropods. These segments include the brain and the main sensory organs. They are usually referred to as the **pre-gnathal segments**, since in mandibulates they lie anterior to the segments that bear the mouthparts, the gnathal segments. In myriapods and insects the three pre-gnathal segments are joined by three gnathal segments to form the head capsule. In crustaceans, additional segments are recruited to these six to form a compound cephalothorax. In chelicerates, the pre-gnathal segments are joined by four segments bearing walking appendages to form the prosoma.

Fossils of Cambrian arthropods indicate that the pre-gnathal segments composed the original arthropod head. In the early evolution of arthropods, the head included only these three segments, and the addition of posterior trunk segments to give the head capsule/cephalothorax/prosoma occurred separately in different lineages. A question remains as to the origin of the pre-gnathal segments, since fossils of even earlier arthropod relatives, as well as extant tardigrades, have a head composed of a single segment. The common view is that the ancient single head segment was joined by two trunk segments to give the three-segment head, in the same way that the three-segment head was joined by trunk segments later in the different lineages. Recent data from comparative embryology suggests that the three-segment head may have evolved by expansion of the single-segment head, and all three pre-gnathal segments are homologous to the original single anterior segment that appeared early in the evolution of Panarthropoda.

CHAPTER 24

Transport and gas exchange systems

Size and transport system

As we saw in Chapter 19, an increase in size bears consequences for many physiological processes. Notably, as size increases, movement of substances through the body by diffusion becomes less efficient. Animals that are larger than a few millimeters in size have to evolve an efficient system for transporting various substances to their targets. These transport systems are needed to carry oxygen and nutrition to the cells and to remove carbon dioxide and waste products away from them. In addition, they carry signals between different tissues of the body in order to allow organism-wide physiological coordination. Transport systems include both circulatory systems, which involve metabolic or physiology-related fluids circulating through the body, usually via distinct vessels, and diffuse systems, which involve a broadly dispersed fluid that reaches all or most of the tissues.

Coeloms as transport systems

We already met coeloms in Chapter 16, where they were introduced as mechanical support systems. Coeloms have an additional role (possibly their ancestral role) as cavities through which various dissolved substances reach the tissues. A fluid-filled coelom is slightly more efficient for transport than relying on diffusion among cells, because the fluid moves and mixes. However, such a coelom is not as efficient as a circulatory system, in which the fluid is pumped actively. Small animals or animals with low levels of metabolic activity may rely exclusively on the coelom for transport.

Circulatory systems

Circulatory systems provide a way for delivering dissolved substance across greater distances and in a more directed fashion. The fluid in the circulatory system may be continuous with the coelomic fluid but it reaches beyond the coelom itself. It is usually referred to as **blood** or as **hemolymph**. Blood is more than just water with dissolved substances. It often includes dedicated proteins that serve as carriers for oxygen, in order to make oxygen transport more efficient. These proteins are known as blood pigments, and they give blood its characteristic color. Different taxa have different blood pigments. Vertebrates use the iron-based protein hemoglobin, and therefore, vertebrate blood is red. Other taxa have copper-based pigments, making their blood greenish or bluish. Blood pigments may be dissolved directly in the blood or concentrated within specialized cells. The red blood cells of vertebrates are an example of such specialized cells.

The blood is circulated through the body with the help of a muscular pumping organ. Such an organ is referred to as a **heart**. As in many other cases we have come across, it's important to remember that although we use the same term for structures in different organisms, this does not mean they are homologous. Hearts have evolved independently numerous times and are as different as the circulatory systems in which they are found. The heart is usually an expansion of a circulatory vessel, with expanded muscle tissue allowing it to contract and push blood through the vessel.

Circulatory systems in which the fluid is contained within vessels for a full cycle from the heart

Organismic Animal Biology. Ariel D. Chipman, Oxford University Press. © Ariel D. Chipman (2024). DOI: 10.1093/oso/9780192893581.003.0024

and back are known as closed circulatory systems (Figure 24.1A). The term "blood" is usually reserved for the fluid circulating in closed systems. In these systems, the vessels split into smaller and smaller vessels, ultimately reaching every cell of the body. In this case, the smallest blood vessels are known as **capillaries**. In vertebrates, the blood vessels leading from the heart are different in shape and structure from the vessels leading to the heart. The vessels leading from the heart to the rest of the body are known as **arteries** and the vessels collecting blood from the body and returning it to the heart are known as **veins**. Other phyla don't always show such a clear distinction between the vessels, but the terms artery and vein are sometimes used outside of vertebrates.

In contrast, open circulatory systems are those in which the fluid is contained in vessels for only part of the cycle, and in the coelom for the remainder of the cycle (Figure 24.1B). In a sense, open circulatory systems are intermediate between a coelom functioning as a transport system and a closed system. In open circulatory systems, hemolymph is pumped by the heart in order to keep it circulating all the time, but there are no vessels leading to all cells. Vessels leading from the heart are known as **efferent** vessels, and vessels leading to the heart are known as **afferent** vessels. The fluid emerges from the efferent vessels into a central cavity, where all the tissues are bathed by the circulating hemolymph. This cavity is a modified coelom and is usually referred to as the **hemocoel**.

(a)

(b)

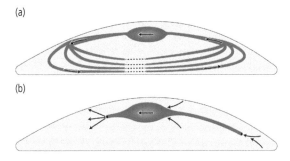

Figure 24.1 Types of circulatory systems: (a) A closed circulatory system, with blood flowing in vessels for the entire circuit. Capillaries are marked with a dotted line. (b) An open circulatory system where the heart pushes hemolymph through a series of tubes into the hemocoel. Arrows indicate the direction of blood/hemolymph flow.

Different roles for circulatory fluids

The main role of blood and hemolymph is to transport substances that are necessary for cellular function to the cells and to remove waste products from the cells. This is probably their original role. However, circulatory fluids have evolved other roles over time. An important function that has evolved in several lineages is wound healing. Blood and hemolymph include various factors, either diffusely in the fluid or packaged in cells, that are activated once they come in contact with air or water outside the body, indicating that the body wall has been wounded. Once activated, these factors come together to achieve clotting of the blood. The blood clot forms a solid impermeable covering over the wound, preventing loss of additional blood and providing valuable time for slower wound healing processes to repair the damaged tissue.

In addition to its role in wound healing, the circulatory system often provides an additional protective function by housing some or all of the **immune system**. The immune system is the collection of all the different defense mechanisms an organism has against invading parasites and pathogens. Some elements of the immune system are innate, in the sense that they have been optimized over evolutionary time to detect and neutralize common pathogens. Other elements can be adaptive, because they learn and adapt throughout an organism's life to react to a specific detrimental agent. The relative importance of innate and adaptive immune responses varies among different taxa.

The circulatory system functions as an internal communication channel, transferring signals between different organs and systems. The signals are usually small diffusible molecules. They can be generated in dedicated glands or cells at different regions of the body. These signals are collectively known as **hormones**, and the system that includes all of the various hormones and hormone-producing glands as well as hormone receptors and hormone target tissues is known as the **endocrine system**. Hormones can convey information about the physiological state of the organism, signaling to distant parts of the body to start or stop producing a specific enzyme or substrate. They can coordinate life history transitions, telling a range of different systems to prepare for sexual maturity or to enter a molting cycle. Hormones can initiate rapid responses to perceived threats, such as the "fight or flight" reaction mediated by adrenaline. In all these examples, the hormone takes advantage of the extent of and the efficient transport by the circulatory system to convey a rapid message to relevant receptive tissues.

As noted above, circulatory fluids are more than just water with dissolved substances. Many of the functions described above are mediated by specific cell types. There are oxygen-carrying cells, cells that aid in clotting, immune cells and other specialized types. In fact, blood and hemolymph should be thought of not just as fluid flowing through the body, but as a tissue in its own right, with typical cell types that undergo differentiation just like cells in any other tissue. The only difference is that rather than being embedded in an extra-cellular matrix, these cells are suspended in a fluid.

Respiration, breathing, and gas exchange

Most metabolic processes in animals require oxygen. Therefore, all cells require a regular supply of oxygen to function. Many metabolic processes generate carbon dioxide. Since carbon dioxide is toxic at high concentrations and leads to acidification, all cells require carbon dioxide to be removed regularly. The terms **respiration**, **breathing**, and **gas exchange** are often used interchangeably for these two processes, although they have slightly different meanings. Respiration usually refers to the physiological process of absorbing oxygen and releasing carbon dioxide, often at the cellular level. Breathing is usually reserved for the process of inhaling and exhaling gaseous oxygen and carbon dioxide. Gas exchange is a general term referring to the movement of oxygen and carbon dioxide to and from tissues and across boundaries. However, these terminological distinctions are not followed consistently in all sources.

Gas exchange is essentially a passive process, with oxygen and carbon dioxide moving by diffusion from areas with high concentration to areas with low concentration. Thus, as carbon dioxide

accumulates in the cell as a result of metabolic processes, it diffuses out of the cell, if there is a lower external concentration. As oxygen is depleted in the cell more oxygen enters it, if there is a higher external concentration. The same is true in gas exchange across surfaces and in circulatory systems.

This process of passive gas exchange can be made more efficient through **counter-current exchange** (Figure 24.2). This involves having gas exchange between two adjacent vessels with blood/hemolymph flowing in opposite directions. The fluid in one vessel at one end of the exchange will have a high concentration of oxygen while the second vessel will have a lower concentration, leading to oxygen diffusing from the first to the second. As fluid flows along the first vessel, it will always find a lower level of oxygen in the second vessel, as the fluid that has already absorbed oxygen has moved away in the opposite direction. The same process happens in the opposite direction with carbon dioxide. Thus, one vessel will enter the exchange with a high concentration of oxygen and a low concentration of carbon dioxide and will leave it after transferring its oxygen to the second vessel and absorbing carbon dioxide from it. The second vessel will do the opposite, beginning with a high concentration of carbon dioxide and a low level of oxygen and ending rich in oxygen and depleted in carbon dioxide.

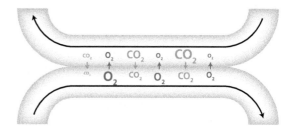

Figure 24.2 Counter current exchange: Two vessels are illustrated with fluid flowing in opposite directions (arrows). The concentration of carbon dioxide (CO_2) decreases in the upper vessel as fluid flows to the left and increases in the lower vessel as fluid flows to the right. At any point of contact, the concentration of carbon dioxide in the upper tube is higher than in the lower tube, causing it to move from the upper to the lower vessel. The situation with oxygen (O_2) is the same but in the opposite orientation.

Gas exchange in water

As we've already seen in Chapter 22, all animals are ancestrally marine. Primitive transport systems are thus adapted for absorbing and carrying oxygen that is dissolved in water. As usual, gas exchange is easiest to understand in very small animals. Oxygen can diffuse freely from the surrounding water and reach all the tissues, supplying them with enough oxygen for their needs. Excess carbon dioxide can similarly diffuse out of the animal into the water. This type of gas exchange is known as **cutaneous respiration**. Many animals in almost all phyla make do with such respiration.

In larger animals, cutaneous respiration cannot supply enough oxygen due to surface area to volume issues. Therefore, different taxa have evolved ways of increasing the surface area through which oxygen can diffuse. This usually happens in discrete locations and not over the entire body surface. Localized extensions of the body wall that increase respiratory surface area are known as **gills**. The gills are usually supplied with blood/hemolymph by dedicated vessels, or via an extension of the hemocoel into the gills. Gills can be distributed along the body (as in the parapodia of annelids), or localized to the anterior of the body, as in fish. The anterior localization helps to first oxygenate the brain, which is often the organ with the highest oxygen demand in the body.

Gills are fragile structures, because they are made of thin tissue to allow gas diffusion through the integuments and because they tend to be made of delicate folds or extensions. They are thus often protected within a semi closed cavity—in this case they are known as **internal gills** (Figure 24.3A), as opposed to the more common **external gills** (Figure 24.3B). Internal gills have an additional advantage in that the gill cavity can have active pumping of fresh water over the gills to make oxygen absorption more efficient.

Breathing in the air

Absorption of oxygen from the air directly into the circulatory fluid is difficult, even though atmospheric oxygen levels tend to be higher than dissolved oxygen levels in water. To be absorbed into the circulatory system, oxygen has to first be dissolved in water that is external to the body. In Chapter 22, we listed desiccation as one of the most significant challenges of living in air. Having exposed wet tissues (e.g., external gills) is a potential source of significant water loss. Therefore, terrestrial animals tend to have internalized gas exchange organs. Air can enter these organs through a narrow, controlled opening that minimizes water loss.

The different breathing organs of terrestrial animals are usually divided into two types (and once again, organs that are given the same name are not always homologous). **Lungs** are single or paired cavities, with a single opening through which air enters and leaves (Figure 24.3C). Lungs are often actively aerated with air entering the lungs and leaving in a cyclical manner. **Tracheae** are blind tubes that may be distributed along the entire body. Air enters the tracheae through openings known as **spiracles** (Figure 24.3D). In many cases, spiracles

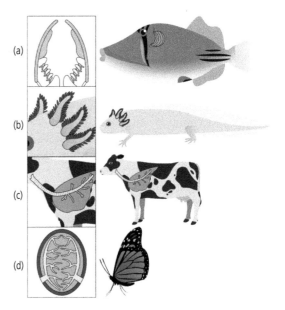

Figure 24.3 Different gas exchange organs: (a) Internal gills in a fish. The gills are enclosed in a cavity and are illustrated as though seen through the body. Their structure can be seen in the cross section. (b) External gills in a salamander. The gills are exposed directly to the water. The increased surface area can be seen in the enlarged image. (c) Lungs in a cow, illustrated as though seen through the body. (d) Tracheae in a butterfly. The spiracles are very small and are distributed along the body. The enlarged cross-section shows the distribution of the tracheae within the body cavity.

have a muscular valve that allows them to open and close in order to minimize water loss. In both tracheae and lungs, the innermost region is lined with a thin epithelium that is covered in water on its apical side. The basal side of the epithelium is in direct contact with the circulatory fluid either within the hemocoel or through capillary vessels.

From an evolutionary point of view, terrestrial gas exchange organs are often modifications of existing breathing organs that have been adjusted by natural selection to function in an aerial environment. Such are the lungs of pulmonate gastropods (a modification of the gill cavity) and book lungs of arachnids (a modified version of the book gills found in horseshoe crabs). In other cases, they are structures that evolved de novo with terrestrialization, such as the tracheae of insects. Vertebrate lungs evolved through an intermediate phase, where the ancestors of terrestrial vertebrates had both a functioning gill system and a primitive gas exchange system via an air sac. Extant lungfish still display this condition.

CHAPTER 25

Embryogenesis

An overview of embryogenesis

Embryogenesis, or embryonic development, is the process in which a single cell develops into a multicellular organism. The animal's body plan is laid out during embryonic development, starting from its very broad organization and refining more and more details as the process of development progresses. Development has been a prime focus of biological research since the mid nineteenth century. In its early days, the study of development focused on the mechanics of tissue movement and differentiation. Experimental approaches that manipulated the developing embryos were the main avenue of research throughout most of the middle of the twentieth century. The molecular revolution of the late twentieth century shifted the focus to genes and tended to describe embryonic development as a series of gene activity and interactions. In recent years, the study of development has taken on a more holistic approach, combining gene activity, with cellular and tissue level analyses, using manipulations that target gene expression at different levels.

Current understanding of development sheds light on the complex relationship between the genome and the organism's morphology. While the information for constructing the organism is encoded in the genome, it is not a straightforward encoding. There are no genes that are responsible for specific body parts or specific shapes. Rather, the genome encodes the machinery that carries out the developmental process—the so called "developmental toolkit" (see Chapter 32 for more on this topic). It also encodes the instructions for where and when to activate the machinery. The ordered and tightly regulated activation of different aspects of development, within the context of the topology of the embryo, results in the gradual assembly of organismal form. In other words, the genome does not directly encode morphology but encodes the process that generates morphology, and that process is embryogenesis.

Embryogenesis can be divided into several phases, which are fairly general and are relevant to nearly all animals. The specifics of each phase tend to be conserved at higher taxonomic levels, but there is also a lot of intra-taxon variability. Because of its importance for laying out the animal's morphology, embryogenesis itself is under very strong selection, but this selection can be both for maintaining a very conserved process and for varying the process in response to different environmental factors. We will cover conservation and variability of developmental processes in Chapter 32, which deals with evolutionary developmental biology. In this chapter, we will cover the general and conserved aspects of embryonic development in animals.

Fertilization

In sexually reproducing species, an individual organism's life begins at the moment of fertilization. In almost all cases, a single **sperm** fertilizes a single egg or **ovum**. The sperm is absorbed by the egg, and the sperm nucleus is shuttled toward the ovum's nucleus, where the two nuclei fuse to form a **zygote**—a single cell with a genome composed of both maternal and paternal contributions. Fertilization of an ovum by more than one sperm cell is prevented by an array of fertilization barriers. The earliest barrier is initiated at the moment the sperm makes contact with the ovum's outer membrane,

Organismic Animal Biology. Ariel D. Chipman, Oxford University Press. © Ariel D. Chipman (2024). DOI: 10.1093/oso/9780192893581.003.0025

when an intracellular cascade of activities changes the properties of the ovum's outer coating to make it inaccessible to subsequent sperm cells. Even if this barrier fails, there are usually additional mechanisms that prevent more than one sperm nucleus from fusing with the ovum's nucleus.

The attachment of the sperm to the ovum is species-specific, as there are receptors on the ovum surface that only bind sperm from the correct species. This recognition is, of course, not perfect, and sperm from closely related species can and do bind to the ovum. Going back to our discussion on speciation in Chapter 2, once enough changes have accumulated between two nascent species so that the ovum of one no longer recognizes the sperm of the other, the two species can be said to be fully reproductively isolated.

Cleavage

Once the ovum is fertilized and the zygote is formed, it starts dividing in order to increase the number of cells, as the first step toward becoming a multicellular organism. This process of cell division is known as **cleavage**. The first rounds of cleavage are usually synchronous, with the number of cells doubling every round, resulting in 2, 4, 8, 16, 32, and so on, cells in successive rounds. The number of synchronous rounds of cleavage varies among different organisms. During the cleavage process, the embryo does not usually increase in size, so the resulting cells are smaller and smaller with each round of cell division. In many cases, the resulting cells are also not morphologically differentiated, so that cleavage ends with a ball of cells that are almost identical in size and shape. In other cases, after three or four rounds, divisions become asymmetric resulting in larger and smaller cells (referred to as **macromeres** and **micromeres** respectively).

Cleavage can be divided into a number of different types, based on a number of parameters. The first major distinction is between **radial** and **spiral cleavage** (Figure 25.1 A, B). The difference between these two is apparent at the third cleavage round when the embryo is composed of eight cells. In radial cleavage, the cells are arranged in a cube, with each layer of four cells sitting exactly above the other four cells. In spiral cleavage, one layer of cells

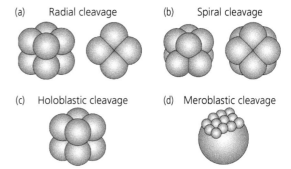

Figure 25.1 Different modes of cleavage.

is rotated by 45° relative to the other layer. While this may seem like a fairly minor difference, there is a much more significant difference between radial cleavage and spiral cleavage. In spiral cleavage, the cells are already committed to different fates by the eight-cell stage. That is, although the cells are morphologically identical, they have already started their process of differentiation, and each cell will give rise to different tissues. In radial cleavage, the cells at this stage haven't undergone any determination yet, and are still developmentally flexible. As we will see below, this difference in cleavage patterns carries phylogenetic significance.

Another distinction is based on the amount of yolk present in the egg or zygote and the degree to which cleavage cuts through this yolk (Figure 25.1 C, D). In **holoblastic cleavage** the egg is split entirely to give rise to separate cells as described above. In **meroblastic cleavage**, the egg is too large and yolk-filled to cleave all the way through, so that cell divisions only take place at one pole of the embryo, leaving an uncleaved yolk mass below it. This difference is not as phylogenetically significant and has more to do with maternal ecology: how much the mother invests in each egg and where she chooses to lay her eggs. We sometimes find holoblastic and meroblastic cleavage in closely related species.

Not all species undergo cleavage as clearly as described here. There are cases where the early stages of embryogenesis are quite different. In some taxa, most notably in arthropods, there is no whole-egg cleavage, and nuclei undergo rapid divisions deep within the yolk, without the egg itself dividing

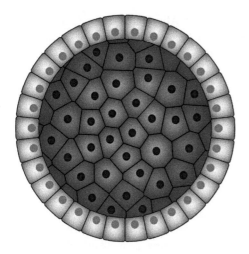

Figure 25.2 Section through a blastula-stage embryo.

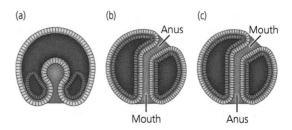

Figure 25.3 Different modes of gastrulation: (a) Section through a schematic embryo at the early stage of gastrulation. Cells have started invaginating through the blastopore (bottom), and the coelomic cavities have started forming laterally to it. (b) Section through a schematic protostome embryo at the end of gastrulation. The blastopore has become the mouth. (c) Section through a schematic deuterostome embryo at the end of gastrulation. The blastopore has become the anus.

into separate cells. This is known as **endolecithal cleavage**.

Toward the end of the cleavage process, when the embryo is composed of a few tens of cells, it is referred to as a **morula**, literally meaning a small berry. Shortly afterward, the cells of the embryo start to concentrate on the outside leaving a fluid-filled space in the middle of the embryo. This stage is known as the **blastula**, and the internal space is known as the **blastocoel** (Figure 25.2). In species with meroblastic cleavage (including many vertebrates with large eggs, such as birds), there is no obvious blastula stage and no distinct blastocoel.

Gastrulation

Up to the late blastula stage, the embryos of most taxa undergo relatively few changes in cell type or identity. In the subsequent process of **gastrulation**, cells start to differentiate and adopt different fates. At gastrulation, the different germ layers are determined, and in many cases, the axes of the embryo are defined (Figure 25.3).

Gastrulation is a complex process of cell movements, in which cells from the exterior of the embryo—known as a **gastrula** at the end of this stage—invaginate into the blastocoel. As cells invaginate, they change from epithelial cells into mesenchymal cells and form the mesoderm. The point on the embryo's surface at which invagination begins is known as the **blastopore**. Gastrulation creates an indentation in the embryo, which gradually expands to form a cavity, and then elongates to form a tube. This is the **archenteron**, which will be the basis for the digestive system. Cells that line the archenteron differentiate into the endoderm. The cells that do not invaginate, and remain on the exterior of the embryo, differentiate into the ectoderm.

At this stage, we can identify two main modes of gastrulation, which are phylogenetically significant. In some taxa, the archenteron elongates until it forms an opening on the opposite side of the embryo. This new opening will become the posterior end of the digestive system, the anus, while the site of the blastopore will form the mouth. Taxa with this type of gastrulation are known as **protostomes**, literally "mouth first." In other taxa the opening that forms from the elongating archenteron will be the mouth, while the site of the blastopore will form the anus. These are known as **deuterostomes**, literally "mouth second."

Organogenesis

Gastrulation lays down the fundamental underlying aspects of the body plan: axes and germ layers. On the basis of this preliminary body plan, embryogenesis then proceeds to define the different tissues and organs, in a process known as **organogenesis**. This process differs significantly among various taxa, although many principles are conserved across all animals. We will give an overview of vertebrate organogenesis, as an example of how the process can work, in Chapter 31 where we discuss the vertebrate body plan.

The phylogenetic significance of developmental modes

Embryogenesis presents an interesting mix of highly conserved processes and processes that change readily across short phylogenetic distances. Of the phases we described above, there are two that are conserved enough that they can be used to make phylogenetic statements. The distinction between deuterostomes and protostomes was observed in the late nineteenth century and used to divide bilaterian animals into two branches. Molecular phylogenies of the late twentieth century and onward, repeatedly confirm this division. The protostome–deuterostome split is the first major division within Bilateria (but see Box 11.2 for a discussion of a possible earlier branching point). However, it is important to point out that as more phyla are studied in detail, the picture that emerges is not as clear cut as it is usually presented, and some protostome phyla turn out to actually have deuterostome-like development. All of the bilaterian phyla we've discussed up to this point have been protostomes. From Chapter 26 to the end of this book, we will discuss deuterostome phyla.

The distinction between spiral and radial cleavage was also made in the late nineteenth century, and like the distinction between deuterostomes and protostomes was suggested to be phylogenetically significant. However, unlike the deuterostome–protostome split, the phylogenetic distinction of cleavage types was not universally accepted, because their distribution among animals is complex. Radial cleavage is by far the more common mode of cleavage, while spiral cleavage is known from a small number of phyla—Mollusca, Annelida, Platyhelminthes, and a few others. Molecular phylogenetics from the early twenty-first century united all of the spirally cleaving phyla, together with a few smaller phyla that display radial cleavage. Detailed analysis of the early development of these smaller phyla suggests that they have probably secondarily evolved radial cleavage, and their ancestral mode of cleavage was spiral. Spiral cleavage can thus be seen as an ancestral character that unites a major branch of the bilaterian evolutionary tree, aptly named Spiralia.

CHAPTER 26

Echinodermata

Introduction to Deuterostomia

As we saw in Chapter 25, the deuterostomes are the second major branch of bilaterian diversity. The diversity of deuterostomes is much lower than that of protostomes, with only three currently existing phyla. Nonetheless, they represent an important group, if only because one of those three phyla is Chordata, the phylum that includes many of the largest and most familiar animals, and to which we ourselves belong.

The main unifying character of Deuterostomia is deuterostomial development, in which the initial site of invagination during gastrulation becomes the anus. Beyond that, many of the developmental characteristics of deuterostomes, such as radial cleavage, are considered to be primitive for Bilateria in general. Indeed, there is recent evidence to suggest that deuterostomial development may itself be primitive for Bilateria, with protostomial development evolving from it. Many deuterostomes have gill slits (a character we will discuss further in Chapter 27), and this may be another ancestral defining character of the group, but there is no consensus on this point.

The echinoderm body plan

While echinoderms ("spiny skin") undoubtedly belong within Bilateria, they have a highly modified body plan, and have lost many of the characters of bilaterians, most notably bilateral symmetry in the adult stage (but some taxa have re-evolved bilateral symmetry). The basic symmetry of echinoderms is pentaradial symmetry (see Chapter 8), although this is clearer in some taxa than in others. The main axis is the oral–aboral axis (similar to what we saw in Cnidaria in Chapter 9). This forms the axis of rotation, around which the five-fold repeated elements are arranged.

Echinoderms have several other unique characteristics that set them apart from all other phyla. Their skeleton is formed of mesoderm-derived calcium carbonate. The skeleton supports the body wall, but it is not a shell and is not external. Since it is covered by a layer of epithelium, it is technically an endoskeleton. The calcitic skeleton is composed of separate plates that in some taxa interlock tightly to form a solid box or **test**, and in some taxa, are small and separate and are embedded in connective tissue, giving a more flexible body wall. These plates have a unique porous microstructure known as **stereom**.

The body wall of echinoderms also contains **mutable collagenous tissue** (sometimes called mutable connective tissue). This echinoderm-specific tissue type is able to change its physical properties under neuronal control and become alternately rigid or relaxed. This ability to change the properties of the collagenous tissue provides an energy-efficient method of maintaining body posture. In addition, it provides functional flexibility that a permanently rigid skeleton does not allow.

The most intriguing echinoderm-specific character is the **water vascular system**. This is a network of water-filled canals and cavities, derived from one of the ancestral coelomic cavities. The system has extensions known as **tube feet** that are mounted externally on pores that cross the skeleton and can reach into the surrounding water (Figure 26.1, Figure 26.2). The water vascular system replaces many systems that are not found in the echinoderm body plan and functions in transport, gas exchange, excretion, food gathering, and locomotion.

Organismic Animal Biology. Ariel D. Chipman, Oxford University Press. © Ariel D. Chipman (2024). DOI: 10.1093/oso/9780192893581.003.0026

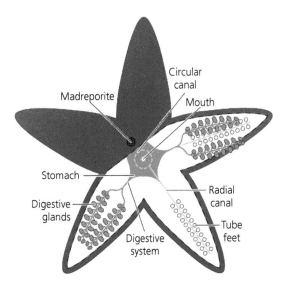

Figure 26.1 Overview of echinoderm body organization: Main echinoderm organ systems illustrated in a starfish, viewed from above. Three of the arms and the central region are seen dissected. One arm shows the digestive system only (bottom left), one shows the water vascular system only (bottom right), and one shows both (top right), with one branch of the digestive system moved to the side. The circular canal is below the stomach and is marked with a dotted line.

Echinoderm diversity

Echinoderm diversity is largely based on the expansion or reduction of different aspects of the body plan (see Figure 26.3). Some echinoderm taxa have evolved novel axes, giving them a new symmetry that is not based on the axes of their pentaradially symmetrical ancestors or of the even more ancient bilaterally symmetrical ancestors. Echinoderms are traditionally divided into five classes. We will list the general characteristics of each class, before discussing their specific adaptations under the different body systems:

Crinoidea (sea lilies)—Sea lilies are suspension feeders with a conservative body plan that is believed to be the closest to that of early echinoderms. The 600 or so species in this class may be stalked or unstalked (the unstalked species are sometimes known as feather stars). The mouth is located at the center of the top of the main body, while the anus also faces upward and lies to one side of the mouth. A ring of feeding arms surrounds the mouth. Juvenile crinoids have five arms and

add arm branches throughout growth, thus making a complex feeding structure that reaches into the water to trap drifting food particles.

Asteroidea (sea stars or starfish)—The scientific name comes from "aster" or star, referring to their structure with five distally tapering arms connecting to central region, without a clear distinction between the arms and the center. Although five arms are the norm, some species have more. The mouth of asteroids faces downward toward the sea bottom, while the anus (which is not always present) faces upward on the aboral side of the animal. Most of the nearly 2000 species of asteroids are predators, many of them being specialized bivalve predators.

Ophiuroidea (brittle stars)—The scientific name of this class literally means "snake tails" and is derived from their thin serpent-like flexible arms. The many-jointed arms connect to a distinct central disk. As in asteroids, the mouth faces downward and is the only opening of the digestive system as no ophiuroids have an anus. This is the most species-rich echinoderm class, with about 2000 described species, most of them predators or scavengers.

Echinoidea (sea urchins)—The typical echinoid is shaped like a somewhat flattened ball covered with calcitic spines. The echinoid **test** is normally made of fused plates and is rigid, usually preserving its shape even after the animal has died. The mouth faces downward, and the anus faces upward opposite the mouth. However, many echinoids stray from this typical globular shape and have an irregular structure, with the mouth and anus displaced from the center. Irregular echinoids (heart urchins, sand dollars) are flattened or discoid and have a clear anterior–posterior axis that is secondarily derived. There are about 1000 species of echinoids, mostly scavengers and algae feeders.

Holothuroidea (sea cucumbers)—By nearly any standard, holothuroids are bizarre animals, with a body plan that is divergent even from the already quite unusual echinoderm body plan. The scientific name derives from the ancient Greek and its origin is obscure, though some believe it comes from "entirely" and "ugly," which is surely an unfair characterization of this fascinating group of echinoderms, some of which can be beautiful colored inhabitants of coral reefs. The English name comes from the fanciful resemblance to the vegetable. Holothuroids have an elongated oral–aboral axis that is functionally an anterior–posterior axis, with the mouth at the front end. Note that this is again a novel axis and is unrelated to the equally novel axis of irregular echinoids. Holothuroids lie on their side with the pentaradially arranged feeding tentacles encircling the mouth. Their skeleton is composed of small, calcite plates embedded loosely in the body wall, which can harden with changes in the mutable collagenous tissue. There are just under 2000 described species in this group.

Echinoderm life history

Nearly all echinoderms reproduce sexually, and with very few exceptions, have separate sexes. Fertilization is always external, with sperm and ova released synchronously into the water column, where fertilization takes place, although there are cases in which the mother retains the ova, which can be fertilized on the animal's surface or in brood pouches. Echinoderms may have direct or indirect development, although indirect development is more common. Each of the classes has a typical larval form. The larvae are bilaterally symmetrical, and the adult body develops from a small rudiment inside the larva.

Echinoderms have extensive regenerative capabilities. Asteroids, crinoids, and ophiuroids can regenerate severed arms. Asteroids can in some cases regenerate the entire body from an isolated arm. Holothuroids will eject their entire digestive system as a predator deterrent and can then develop a new digestive system in its place. Even echinoids have limited regenerative abilities and can replace broken spines or even regrow damaged plates. As we saw in annelids, this type of extensive regenerative capacity allows asexual reproduction by fission, and this is indeed found in a few echinoderm taxa.

The water vascular system

The water vascular system is unlike anything in any other phylum. Despite its seemingly simple structure, it is a highly coordinated system with numerous functions. The water vascular system is derived from the coelom and is thus a fluid-filled series of tubular cavities that are separate from the main body cavity. At the basis of the water vascular system is a circular canal that lies around the mouth opening. From this canal, five radial canals extend into the more distal parts of the body. The circular canal is connected via the **stone canal** to a specialized skeletal plate known as a sieve plate or **madreporite** (Figure 26.2). The sieve plate allows surrounding sea water into the stone canal but prevents particulate matter from entering. The fluid inside the water vascular system is thus continuous with the surrounding sea water.

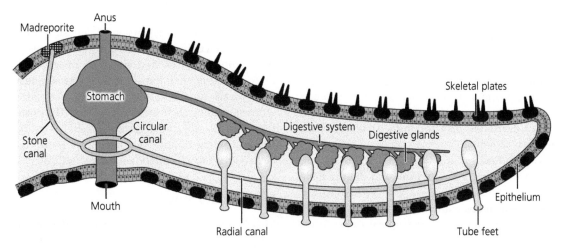

Figure 26.2 Echinoderm internal structure: Main organ systems of echinoderms illustrated with a longitudinal section through an arm and the central region of a starfish.

Epithelia-lined protrusions, the tube feet, connect to the radial canals via lateral canals, and extend out into the surrounding water (Figure 26.2). In asteroids, holothuroids, and echinoids, ampullae at the base of each of the tube feet push water into the tube feet, causing them to extend. Muscles along their shafts allow the tube feet to retract when the ampullae ease water pressure. In some echinoderms, the tube feet terminate in suckers, and coordinated movement of numerous tube feet aids locomotion. In other cases, the tube feet serve to move food toward the mouth. Tube feet can also act as chemoreceptors, swaying in the water and picking up chemical scents. Tube feet are also like gills in mediating gas exchange and they provide a surface through which excretion can take place.

The water vascular system of crinoids and holothuroids is modified relative to this description. Crinoids have no madreporite, and the fluid inside the system is separate from the environment. The tube feet have no suckers and no role in locomotion, and function mainly to transfer food to the arms and as chemosensory organs. Holothuroids tend to have a reduced water vascular system, but the degree of reduction varies among different groups. The tube feet are modified to form oral tentacles that surround the mouth. Some holothuroids have no tube feet along the body, whereas others have them only along the "ventral" surface. There is some evidence to suggest that holothuroid tube feet evolved convergently to the tube feet of other echinoderms.

Feeding and digestion

Feeding modes in echinoderms include suspension feeding in crinoids, deposit feeding in echinoids and holothuroids, and predation in most ophiuroids and asteroids. The diverse feeding modes are reflected both in their body plan and in the structure of the digestive system.

Crinoids have an upward-facing mouth and use their arms to trap food particles and deliver them to the mouth via mucus-lined grooves in the arms. The intestine loops within through the body cavity ending in an anus that points upward next to the mouth.

The vast majority of asteroids feed by everting their stomach onto food-bearing surfaces for deposit feeding. Their stomach can thus engulf food items and start digesting them before the stomach is brought back inside the body to continue the digestive process. In asteroids that are bivalve specialists, the stomach can be inserted into a thin crack between the valves to start digesting the bivalve's closure muscles, thus weakening the prey's resistance. Some asteroids are suspension feeders, and don't have eversible stomachs.

Ophiuroids have a simple digestive system with a large undifferentiated stomach. Most ophiuroids are predators or opportunistic scavengers. One group of ophiuroids, the basket stars, have highly branched arms that can curve in different directions, forming the eponymous basket. They are suspension feeders that hold onto a prominent rock with the mouth facing downward, raising their many-branched arm trips trap food particles, but also to catch larger prey, which they bring to their mouths with actions of the arms and tube feet.

Echinoids are mostly deposit feeders, herbivores and detritivores and graze the sea bottom as they move along it. Predatory echinoids are rare. Regular echinoids have five calcitic teeth held and operated by a complex feeding apparatus, known as **Aristotle's lantern**, which is used for scraping or biting. The intestine loops within the test and ends in an aboral anus. In irregular echinoids, the mouth is displaced forward, with the Aristotle's lantern lost or highly reduced, and the anus is sometimes strongly displaced backward to form a secondary anterior–posterior axis.

Holothuroids are suspension or deposit feeders. Suspension-feeding holothuroids trap food using the feeding tentacles that surround their mouth. Deposit-feeding holothuroids often eat sediment and extract nutrition from it (much like earthworms), passing the excess to the anus. The posterior of the holothuroid digestive system has a large rectum leading to anus. The rectum is sometimes large enough to accommodate commensal fish.

Muscles and movement

Muscles in echinoderms have a very different role in movement and locomotion than what we've seen in other phyla. The muscles are usually simple epithelial muscles that are often arranged into tubes.

Unlike most other examples covered in previous chapters, muscles in the body wall usually connect the plates, but are also used to change the overall shape of the animal's body wall. They can also activate levers in the form of rings of musculature around the bases of spines. Their role in locomotion is mediated by tube feet, but their role in maintaining body shape is augmented by the stiffing and softening of mutable collagenous tissues, as we have already seen.

Mutable collagenous tissue is very energy efficient since it allows rigidity with no muscular effort. The principle of this tissue is that after muscles change the shape of an echinoderm, the collagen in the tissue "catches" in position and maintains this shape without further energy input. Mutable collagenous tissue makes up most of the body wall in echinoderms. The movement of crinoid and ophiuroid arms is mostly driven by muscular activity, flexing the joints between individual plates. Locomotion is rarely by swimming in which rhythmic movement of the arms push the animal through the water column, but usually by crawling with the arms pushing against the substrate. In holothuroids, locomotion is by crawling along the substrate, by means of musculature and coordinated stiffening and flexing of the mutable collagenous tissue in the body wall, which is strengthened by small embedded calcitic plates.

Tube feet are the main agent for locomotion in asteroids and echinoids, although in the latter, spines also play a major role in many taxa. Muscles in the ampullae internal to the tube feet control water pressure in the tube feet. When they contract, they force water from the ampullae into the tube feet, extending them. Muscles along the tube feet then retract, usually in a coordinated fashion to direct flexing of the tube feet to aid in locomotion. Echinoids and asteroids advance by movement of numerous tube feet. When seen in motion, these animals superficially seem to glide across surfaces, since there is no obvious change in shape of the rest of the body. Echinoids and asteroids have no functional anterior end and can move in any direction. Nonetheless, asteroids are reported to have a preferred "leading arm" and will tend to move in its direction.

In addition to locomotion via tube feet, some echinoids use their spines as "stilts." In this case, the movement of the spines, which are connected via ball-and-socket joints to the calcitic plates, is through muscular activity. Mutable collagenous tissues can be used to hold a spine in a specific orientation, again without use of muscular energy, jamming the echinoid into crevices for long periods with great efficiency. Irregular echinoids use short spines to burrow in soft sediment with coordinated waves of spine movement.

Integument

As already described, the body wall of echinoderms is composed of skeletal plates embedded within varying amounts of mutable collagenous tissue. These plates are mostly composed of calcite that is sometimes strengthened by small amounts of magnesium. In echinoids, the plates are large and held in rigid casings by collagen that microscopically "sews" them together. This is also true for some types of crinoids, asteroids, and ophiuroids, but in most of these, the plates are much more loosely connected together to allow movement. In most holothuroids, the plates are small and scattered within the body wall without being tied together. Outside this body wall is a soft, thin epithelium.

Various types of spines, which are also covered in the same epithelium that covers the rest of the body, serve different functions. In echinoids, these can be not only locomotory, but can also be protective, and are occasionally tipped with protective venom. Echinoids and asteroids have strange, pincer-like structures known as **pedicellariae** scattered over the body. These are thought to have evolved from spines, but they have probably evolved independently in the two groups, as they differ in many ways. The pedicellariae are mostly used to ward off parasites or settling organisms looking for a good place to attach, and for the removal of debris or bacteria ("grooming"). Some of them are large enough to be used against predators and can be equipped with venom. There are a few species of starfish in which pedicellariae are used for prey capture.

Nervous and sensory systems

The echinoderm central nervous system is a simple nerve ring surrounding the mouth. There is no

Figure 26.3 Examples of echinoderm diversity: (a) An asteroid, the elegant sea star *Fromia nodosa*. (b) A crinoid, the feather star *Comanthus*. (c) A holothuroid, the California sea cucumber *Parastichopus californicus*. (d) An echinoid, the purple sea urchin *Paracentrotus lividus*.

Source: a: Photo supplied from Shutterstock: aquapix (https://www.shutterstock.com/image-photo/noduled-sea-star-fromia-nodosa-underwater-757914562); b: Photo supplied from Shutterstock: Bildagentur Zoonar GmbH (https://www.shutterstock.com/imagephoto/feather-star-comanthus-sp-1009076125); c: Photo supplied from Shutterstock: Jaya Kesavan (https://www.shutterstock.com/image-photo/california-sea-cucumber-giant-parastichopus-californicus-1307844247); d: Photo supplied from Shutterstock: Al Carrera (https://www.shutterstock.com/image-photo/sea-urchin-paracentrotus-lividus-cabo-copepuntas-1933745786).

centralized brain and the nervous system functions as a distributed nerve net, which also includes a layer of nerve cells below the external epithelium of the body. Sense organs are mostly epithelial receptors with no clearly distinct complex sensory structures. Echinoderms have no eyes, no mechanosensory antennae, and no external chemosensory organs. Nonetheless, they do respond to chemical cues, to touch, and to water currents. Echinoderms appear to have distributed photoreceptor cells, mostly on the tube feet, that respond to different angles and intensities of light, allowing a rudimentary form of image-detecting vision.

The arms of crinoids and ophiuroids have propriosensory receptors that allow them to regulate the movement of the arms with precision. Several types of echinoderms have statocysts that allow them to maintain the correct orientation on the sea floor, most notably the sea urchins in which this function is performed by structures called **sphaeridia** which, like pedicellariae, seem to have evolved from spines.

Although the nervous system of echinoderms is much less complex than that of many other animals of equivalent size, recall that it allows for the coordination of numerous moving tube feet, for the control of the very active, tiny pedicellariae that can be coordinated in their attacks, and for the flowing movement of the feeding arms of sea lilies and basket stars. Echinoderms are slow moving animals but hidden within that slowness may be some surprisingly intricate behaviors.

Evolutionary history

Echinoderms have a rich fossil record, due to their mineral-based body wall that readily fossilizes. Fossil echinoderms are known from the lower Cambrian (about 520 million years ago), although some Ediacaran fossils (over 550 million years old) have been controversially suggested to represent even earlier echinoderms. Typical plate structure and the presence of stereom identify the Cambrian forms as echinoderms, or at least as close echinoderm relatives. While all modern, and many Cambrian echinoderms are pentaradial, their phylogenetic position indicates that beyond doubt, the phylum itself shares common ancestry with bilaterally symmetrical deuterostomes. Fossil forms show a range of symmetries and shapes, and it is difficult to reconstruct the relationships among the disparate forms and trace a sequence of evolutionary transformations between the diverse fossil body plans. A few of the Paleozoic forms present some level of modified bilateral symmetry and are considered by many workers to represented fossilized examples of the transition from ancestral bilaterality to pentaradiality that characterizes the phylum. Other studies of these fossils suggest that they may be derived from pentaradial ancestors. If this interpretation is correct, it would raise the possibility that the pentaradial body plan evolved very early in echinoderm evolutionary history, possibly even before the earliest known fossils of this phylum.

Echinoderms were significantly hit by most major mass extinctions, and several of the early fossil groups have no living representatives. The forms with unusual symmetries disappeared before the end-Permian mass extinction (see Box 26.1) leaving only the pentaradial forms that we now see. All of the modern classes were severely affected by the end-Permian extinction, leaving only a few taxa that survived to the Mesozoic era. In many cases, these descendants have characteristics quite different from those of their predecessors that were killed off during the mass extinction.

Box 26.1 The end-Permian mass extinction

Extinction is part of evolution. All species go extinct eventually, and throughout the history of life on Earth, there has always been "background extinction." The rate of extinction varies over time, and there are periods with increased levels of extinction, significantly above background levels. These periods of increased extinction rates are known as extinction events or **mass extinctions**. Among the mass extinction events, there are five that are considered to be significant, and that led to dramatic changes in the biosphere. The greatest of the five great mass extinctions is the one that occurred at the end of the Permian and defines the border between the Paleozoic and the Mesozoic Eras.

The end-Permian mass extinction (also known as the Permo-Triassic event) occurred about 252 million years ago. During this extinction, an estimated 95% of all animal species on Earth disappeared, and many higher taxa that had been around for tens or hundreds of millions of years were wiped out. The extinction event was a drawn-out process, lasting tens of thousands of years and probably including a number of distinct pulses, separated by short periods of recovery.

The cause of the end-Permian mass extinction is believed to be a cascade of dramatic environmental changes that led to a runaway process of climatic catastrophes and ecosystem collapse. The initial trigger seems to have been widespread volcanic activity in the area that is now Siberia. Unprecedented amounts of lava poured out of cracks in the Earth's crust, together with carbon dioxide and other gases. The carbon dioxide led to a greenhouse effect, which raised the average temperature dramatically. This led to the release of methane dissolved in the oceans, further contributing to the greenhouse effect. Oceans became acidic, forests dried out, sea levels rose and almost all possible environments on the planet were negatively impacted. The effects were so severe, that the event is known as "the Great Dying" since life on Earth was very close to disappearing altogether.

In the aftermath of the extinction, it took close to 30 million years for environments to return to pre-extinction diversity levels. The world that emerged from the extinction was completely different from what it had been. New types of animals and plants emerged. The new world saw dinosaurs on land and giant reptiles and ammonites in the sea. Seed plants diversified and shortly afterward, flowering plants. With the flowering plants came insect pollinators. The end-Permian mass extinction was a devastating event, but the world recovered. The recovery paved the way for the evolution of the biosphere we live in today.

The exquisite fossil record of the phylum makes them excellent subjects for studies of evolutionary processes. Coupled with newer morphological and molecular techniques, the origins and diversification of evolutionary novelty within the group attract many biologists interested in these subjects. Echinoderms are common fossils throughout the Paleozoic. However, holothuroid fossils tend to be rare because they largely lack significant calcitic skeleton. Fully articulated echinoids, both regular and irregular, as well as a few asteroid and ophiuroid taxa, are found in many deposits, but stalked crinoids are among the most common fossils since the early Ordovician. Isolated elements from crinoid stalks are found in the majority of marine deposits, sometimes making up almost the entire rock mass, and "sea lily gardens" are an iconic component of Carboniferous to Jurassic fossil sites worldwide.

CHAPTER 27

Chordata and Hemichordata

Introduction to Chordata

We now come to the chordates, the last phylum we will discuss in detail. Because of their evolutionary significance and importance to biological research in general a total of four chapters will be devoted to Chordata. Much of chordate diversity lies within Vertebrata, so three of the four chapters (Chapters 29–31) will be devoted to that group. In this chapter we will introduce the general characteristics of the phylum and cover the "invertebrate" chordates. We will also briefly introduce Hemichordata, a third deuterostome phylum that shares some chordate characteristics. Hemichordates were traditionally allied with chordates, but are now recognized as a sister group to the echinoderms within a deuterostome clade known as Ambulacraria. While hemichordates do not technically belong in the same chapter as chordates, we discuss them here as a comparison with chordates, because they shed some light on the shared evolutionary history of the two phyla.

The chordate body plan

The chordates have a series of unifying characters (Figure 27.1). The first of these, which gives

the phylum its name, is the presence of a semi-rigid longitudinal dorsal structure known as the dorsal chord or **notochord**. The notochord is the original internal skeletal support structure of chordates, although it is modified in many taxa and is often only evident in embryos or larvae. Ventral to this structure is the hollow dorsal nerve cord (sometimes called the **neural tube**). Note the possible confusion between the dorsal chord (support structure) and nerve cord (neural structure). The main component of the central nervous system in chordates is dorsal and this is in contrast with all other phyla we've discussed so far, where the nervous system is ventral.

Chordates are also characterized by the presence of **gill slits** or **pharyngeal slits**. Ancestrally, these were part of a feeding and gas exchange system that consisted of an enlarged pharynx which was used to ingest water. The water is pushed out through the pharyngeal slits, filtering out food particles that remained within the pharynx by trapping them with mucus. The filtered particles and mucus are transferred to the digestive tract. This mode of suspension feeding is still found in most invertebrate chordates (and in hemichordates). In vertebrates, the role of these structures has changed

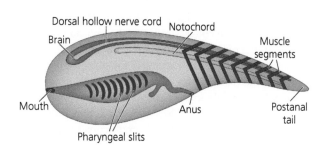

Figure 27.1 Chordate characteristics, illustrated on a generalized chordate.

Organismic Animal Biology. Ariel D. Chipman, Oxford University Press. © Ariel D. Chipman (2024). DOI: 10.1093/oso/9780192893581.003.0027

somewhat and they've been modified to function as a breathing system, with the water being pushed out the pharyngeal slits across epithelial gills. This role remains in all primitively aquatic vertebrates (the diverse assemble we think of as "fish"). In terrestrial vertebrates the pharyngeal slits are a transient embryonic structure, but the support elements of the slits, the so-called **pharyngeal arches**, adopt a range of shapes and functions, as we will see in subsequent chapters.

The chordate body is composed of an anterior–posterior series of chevron-shaped muscles. These muscles extend past the anus to give a post-anal tail. Since the accepted definition of a tail is a structure that extends past the anus, chordates are the only animals that have a real tail in the adult stage.

Hemichordate body plan and diversity

We now make a slight digression to discuss hemichordates, before continuing with a more detailed description of chordate diversity and structure. The prefix "hemi" means half. The hemichordates' name comes from the fact that they present only some of the chordate characteristics. The relative phylogenetic positions of chordates and hemichordates suggest that the characters they share are probably ancestral for Deuterostomia in general. While hemichordates usually receive little attention in textbooks and courses, understanding them—and especially their early development—is crucial to understanding the early evolution of our own phylum.

The phylum Hemichordata consists of close to 150 described species of marine worms, but there are new species being described, especially in deep sea explorations. Most members of the phylum are bottom dwellers or burrowers. The hemichordate body is divided into three regions: a proboscis (prosome), a collar (mesosome), and a trunk (metasome). Some hemichordates have a dorsal nerve cord located in the collar, similar to the chordates. The remainder of the nervous system is mostly a diffuse nerve net. Anterior pharyngeal gill slits or gill pores line the trunk, at least in its anterior region. This is where the obvious similarity with the chordates ends, although there are some hints that the collar includes elements that are homologous to the chordate notochord.

Hemichordates are divided into two classes. The morphology of the two classes is so different that it is not immediately obvious that they are related:

Enteropneusta (acorn worms)—These are benthic, worm-like, mostly burrowing animals (see Figure 27.2). The majority of hemichordate species belong to this group. The proboscis is large and looks like an acorn in some species, giving them their common name. Acorn worms are deposit feeders or detritivores that usually live within a U-shaped burrow. The digestive system is straight and simple. The nervous system in the trunk includes both a dorsal and a ventral nerve cord, in addition to the epithelial nerve net.

Pterobranchia—These are small, suspension-feeding colonial animals with short bodies. Despite their small size, the division of the body into three regions is still evident. Like many small suspension feeders, they often have feeding tentacles. In overall morphology and habitat, they are superficially similar to colonial bryozoans and were originally identified as such. Many structures are reduced due to their small size, and many details of their anatomy are not known. Floating colonies of pterobranchs are preserved in the fossil record throughout most of the Paleozoic and are known as graptolites.

Chordate diversity

The chordates are divided into three subphyla: Cephalochordata (head chordates—the lancelets), Tunicata (sometimes also known as Urochordata), and Vertebrata (see Figure 27.2). Until the early twenty-first century, cephalochordates and chordates were believed to be sister groups, to the exclusion of tunicates. This belief was based on the similarity of the cephalochordate body plan to the hypothesized vertebrate ancestor. Various sources of information, including phylogenomic studies, now point strongly to the fact that it is in fact tunicates that are more closely related to chordates, with cephalochordates belonging to an earlier branching taxon. We only discuss tunicates

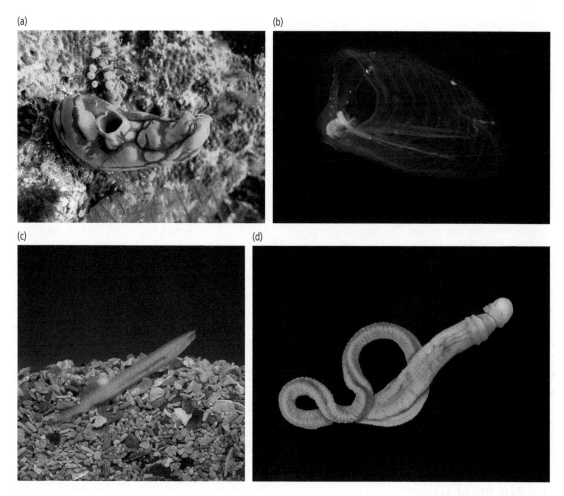

Figure 27.2 Examples of chordate and hemichordate diversity: (a) An ascidian tunicate, *Polycarpa aurata*. (b) A thaliacean tunicate, the salp *Salpa aspera*. (c) A cephalochordate, the lancelet *Branchiostoma lanceolatum*. (d) A hemichordate, the acorn worm *Glossobalanus sarniensis*.

Source: a: Shutterstock, https://www.shutterstock.com/image-photo/tunicate-polycarpa-aurata-raja-ampat-west-1634473249; b: Alamy, https://www.alamy.com/stock-photo-this-pelagic-tunicate-or-salp-salpa-aspera-is-part-of-the-salpidae-18743631.html?imageid=7757775F-F4C9-4191-992A-9A935240116B&p=8797&pn=1&searchId=55a2b9c458dc994d073a3d3f66dc05e1&searchtype=0; c: Alamy, https://www.alamy.com/stock-photo-lancelet-or-amphioxus-branchiostoma-lanceolatum-is-a-marine-cephalochordate-20816420.html?imageid=C13EEF0B-3BF3-47E7-B246-0B7D495B7362&p=6316&pn=1&searchId=c4601004f17845d557ec603737d72cf7&searchtype=0; d: Alamy, https://www.alamy.com/stock-photo-glossobalanus-sarniensis-is-an-acorn-worm-belonging-to-a-group-of-80450002.html?imageid=D6EF8B29-7DD6-4EB9-A1A1-C9347A9743B9&p=14455&pn=1&searchId=9f2fa7fa7afecb04ba3c614d031d0041&searchtype=0.

and cephalochordates here, leaving the discussion of the vertebrates to the chapters dedicated to them.

Cephalochordata is a small subphylum, probably consisting of only 20–30 described species. They are the most "typical" chordates, in the sense that they maintain all chordate characteristics throughout their entire life. Cephalochordates are small, fish-like animals with a translucent body and a small head that is not clearly distinct from the body. They are mostly detritivores or suspension feeders, with

the mode of feeding sometimes changing throughout their lives. They are called lancelets because they reminded early researchers of small dissection lancets. The best studied cephalochordate is *Branchiostoma*, commonly known as amphioxus. Note that despite the Greek-sounding name, this is not its official scientific name, so it is not capitalized or written in italics. Amphioxus has been used for over a century as a proxy for the body plan of the earliest chordates. It is, however, important to remember, that despite its relatively conservative body plan, it has had hundreds of millions of years of independent evolution and has diverged in several subtle but important aspects from its ancient ancestors, so such proxy studies should be done with caution.

Tunicates are much more diverse than cephalochordates, both in species number and in morphology. The 3000 or so described tunicate species can be sessile or planktonic. Some are solitary and some live in colonies. Regardless of their lifestyle, almost all are suspension feeders, using their pharynx and pharyngeal slits to filter food particles and absorb oxygen from the sea water. The largest group of tunicates is Ascidia. Ascidians are sessile suspension feeders that can be solitary or colonial. They have a sac-like body with no tail and a highly reduced head. Water is pumped via an incurrent and an excurrent siphon and filtered by the pharynx. Members of Larvacea retain their larval tail as adults and live as tiny free-swimming tadpole-like animals. They form mucus baskets, in which they live, that allow them to filter large volumes of sea water efficiently. Thaliacea (salps and their relatives) includes species that form long floating or planktonic colonies, with numerous barrel-shaped zooids that can also filter large amounts of sea water in a short time.

Chordate life history

Both cephalochordates and tunicates usually display indirect development. Cephalochordates undergo metamorphosis that involves the planktonic larva settling and becoming endobenthic. While there are many changes in organs that are restructured during metamorphosis, the most dramatic change is a transition from an asymmetric larva to a (mostly) symmetric juvenile. Reproduction in cephalochordates is always sexual, with separate sexes and gametes that are released into the water for external fertilization.

Most tunicates are hermaphrodites. In some cases (most notably in larvaceans) gametes emerge from the animals by rupturing of the body wall, leading to their death. Tunicates normally have a tailed larva that is lost in metamorphosis. In Ascidia, metamorphosis also includes a transition from a planktonic lifestyle to a sessile one. However, some ascidians display direct development. In these cases, large yolky eggs can be produced and a miniature adult hatches from the egg. While a tailed larva is the most common type of larva in ascidians, there are also cases of tailless larvae. Interestingly, tailed and tailless larvae are found in closely related species, with several cases of switching back and forth between the two forms. Larvaceans maintain their tail as adults and are thus considered to be permanent larvae—hence the name.

Colonial ascidians have asexual reproduction in addition to the common sexual reproduction. A founder zooid settles and starts feeding, and grows additional zooids through budding, giving rise to a clonal colony. Similarly, thaliaceans form long, floating colonies, composed of zooids that bud off from a founder zooid. You will recall that very similar types of colony formation are found in many cnidarians (Chapter 9). This is of course convergent. As with many animals that reproduce via budding, colonial ascidians have significant regenerative capabilities. Experiments have shown that they are capable of regenerating a complete colony from small fragments formed by budding.

Feeding and digestion

All cephalochordates are suspension feeders. They usually lie half covered by the substrate, with only their mouth and gill slits protruding above the surface. The mouth is surrounded by tentacles that act as a rough filter and aid in bringing suitable food particles toward the mouth. Water is pumped into the mouth by a ciliary pump. Water is then ejected through the gill slits, with the food remaining inside. A ventral organ known as the **endostyle** produces mucus, which leads food into the digestive system. The digestive system is a simple tube ending in the anus, which is anterior to the posteriormost point of the body. A diverticulum in the gut may be homologous to the liver and pancreas in vertebrates and may thus represent the precursor to these two organs.

Tunicates are also suspension feeders, although their mechanism of food capture is somewhat different from that of cephalochordates and varies among the different taxa. The pharynx is significantly enlarged to form a pharyngeal chamber that is connected to the environment via an incurrent syphon. Ciliary action brings water into the pharyngeal chamber. Water is pushed out via the pharyngeal slits into an **atrium** that is external to the pharynx, but surrounded by an outer cover known as a tunic, which gives the group its common name. The water is ejected from the tunic via an excurrent syphon. As in cephalochordates, there is mucus-producing endostyle. The mucus traps the food particles and is ingested together with them. The digestive tract is U-shaped, with an anus that empties into the atrium, next to the excurrent syphon. Feeding in larvaceans is somewhat different from the description above, as they use a large external mucus net to trap food particles.

Muscles and movement

Cephalochordates and larval tunicates have segmental muscles along the body and tail and move by serpentine swimming motions. The notochord provides the main structural support for this muscle action (Figure 27.3). Most tunicates lose their tail during metamorphosis and become sessile or planktonic. Ascidians settle and adhere to hard substrates using a sticky adhesive substance. Thaliaceans and larvaceans do not settle and remain planktonic throughout their lives.

Integument

Cephalochordates have a simple, single-layered epithelium as their only integument. Tunicates also have a simple epithelium as their main body covering, but they also have an additional external layer, the eponymous tunic. The tunic can be of varying thickness and physical characteristics. It is sometimes thin and transparent and sometimes thick and leathery but always includes cellulose and other polysaccharides as a component, with various additional proteins. When the tunic is thick, it functions as a protective exoskeleton.

Nervous and sensory systems

Cephalochordates have a dorsal nerve cord with minimal anterior expansion. They thus have no

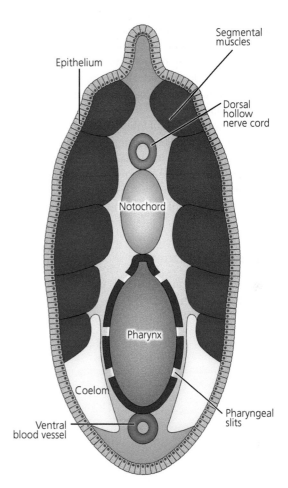

Figure 27.3 Transverse section through a generalized cephalochordate.

Source: a: Photo supplied from Shutterstock: Rocky Cranenburgh (https://www.shutterstock.com/image-photo/tunicate-polycarpa-aurata-raja-ampat-west-1634473249); b: Photo supplied from Alamy: David Fleetham (https://www.alamy.com/stock-photo-this-pelagic-tunicate-or-salp-salpa-aspera-is-part-of-the-salpidae-18743631.html?imageid=7757775F-F4C9-4191-992A-9A935240116B&p=8797&pn=1&searchId=55a2b9c458dc994d073a3d3f66dc05e1&searchtype=0); c: Photo supplied from Alamy: Natural Visions (https://www.alamy.com/stock-photo-lancelet-or-amphioxus-branchiostoma-lanceolatum-is-a-marine-cephalochordate-20816420.html?imageid=C13EEF0B-3BF3-47E7-B246-0B7D495B7362&p=6316&pn=1&searchId=c4601004f17845d557ec603737d72cf7&searchtype=0); d: Photo supplied from Alamy: Nature Photographers Ltd (https://www.alamy.com/stock-photo-glossobalanus-sarniensis-is-an-acorn-worm-belonging-to-a-group-of-80450002.html?imageid=D6EF8B29-7DD6-4EB9-A1A1-C9347A9743B9&p=14455&pn=1&searchId=9f2fa7fa7afecb04ba3c614d031d0041&searchtype=0).

real brain, and this is one of the main things that differentiate them from vertebrates. Segmentally repeated neural extensions from the dorsal cord innervate the muscles and other parts of the body.

Some species have an anterior unpaired pigment spot, which functions as the main visual organ. In addition to this, there are a number of epithelial sensory cells distributed along the body, which function both as tactile receptors and as chemosensory structures.

The tunicate CNS is highly reduced. In the larvae, there are about 300 cells that form a network of interactions. There is no dorsal nerve cord in adult tunicates, but there are several ganglia with specific functions. Beyond that, relatively little is known about the adult tunicate nervous system or sense organs.

Evolutionary history

Fossil chordates are rare but are known from various Cambrian sites. Perhaps the best-known fossil chordate is *Pikaia* from the Burgess Shale. Its fame comes from the fact that it was discussed by Stephen J. Gould in his influential book *Wonderful Life*, where he mused about the hypothetical possibility of *Pikaia* going extinct with no descendants, which would end vertebrate history before it began. We now know that *Pikaia* was probably not a direct vertebrate ancestor, but its fame remains.

Tunicates are almost unknown from the early fossil record. However, in recent years, several animals from the Cambrian that could not be assigned to any known phyla (so called "weird wonders") have been reinterpreted as early tunicates. The lack of any mineralized body parts makes them extremely rare as fossils beyond the Cambrian.

Vertebrates of course have an extensive fossil record, but we will discuss that separately.

CHAPTER 28

Excretory systems

Introduction to excretory systems

Up to now, we have discussed almost all of the organ systems in animals. The last remaining system we have to cover is the excretory system. We have left this to last, because it integrates several other systems, including the digestive system and the circulatory system, and because its function is closely linked to whether the animal lives in fresh water, sea water, or air.

The excretory system has two separate roles, but these two roles always appear together. The first role is the elimination of waste products. All metabolic processes produce unusable byproducts. In many cases, these products are potentially toxic and can cause damage if allowed to accumulate. Therefore, there is strong selection to evolve a method to store this waste where it cannot cause harm, and to excrete it periodically or continuously from the body.

The second role is maintaining a correct ion balance, a process known as **osmoregulation**. This involves regulating the ratio of water to ions, which can be done either by removing water, or by removing ions, or by some combination of the two.

Waste products

When food is ingested and digested it is broken down to constituent units—simple sugars, amino acids, nucleotides, fatty acids, etc. These constituents are then reused to build the molecules the organism needs—new proteins, new nucleic acids, new polysaccharides, and others. Sometimes the simple organic molecules are broken down even further. Sugars and fatty acids are broken down for energy, creating water and carbon dioxide.

Excess carbon dioxide is removed via the circulatory system to the gas exchange system, as we've already discussed in Chapter 24. Amino acids and other molecules with amine groups produce ammonia (NH_4) when broken down. Ammonia is toxic and is also very soluble in water, allowing it to diffuse through the body. In small marine organisms, ammonia diffuses through the body, but also diffuses out, because its concentration in the surrounding sea water is very low. Larger organisms run the risk of accumulating ammonia in internal tissues, and therefore have to actively remove or detoxify it as quickly as possible before it can cause damage. Many animals, most notably vertebrates, convert ammonia to urea, which is a nontoxic soluble compound. Urea can then be collected by the excretory system and removed. Other animals convert ammonia to uric acid, which is less soluble and forms crystals. These crystals can also be removed, usually via the digestive system. Conversion of ammonia to uric acid is common among many terrestrial invertebrates. Vertebrates also generate uric acid as a byproduct. Accumulation of uric acid crystals in the joints leads to the condition known as gout, and uric acid crystals are one of the sources of kidney stones.

Osmoregulation

The body fluids of marine organisms (including coelomic fluids and the fluid in the circulatory system) are very similar in composition to the surrounding sea water. This is known as being **isotonic** with the environment. In the simplest cases, there is a connection between the organismic fluids and the sea water either via the integument or through the circulatory system. Echinoderms (chapter 26) are an extreme example of such a connection, with the

Organismic Animal Biology. Ariel D. Chipman, Oxford University Press. © Ariel D. Chipman (2024). DOI: 10.1093/oso/9780192893581.003.0028

madreporite allowing sea water to enter the water vascular system. While such a connection makes the activity of the excretory system easier, fine tuning of ion balance is still required, especially in larger organisms. Such a direct connection also carries a price, in making the animal more susceptible to infection by pathogens and parasites.

Animals exposed to varying salinity levels—those living in estuaries where fresh water and sea water mix, or those living in tidal zones or in ephemeral bodies of water, where evaporation can change the concentration of ions in the water—are faced with a greater challenge than those living in a stable environment. These animals can evolve to be **osmoconformers** or **osmoregulators**. Osmoconformers remain isotonic with the environment by maintaining a connection with the water around them and can tolerate changes in salinity of their internal fluids. In contrast, osmoregulators isolate themselves from the environment and maintain a constant internal ion concentration regardless of the surrounding salinity. This requires a more active excretory system that invests energy in pumping ions in and out of the organism's tissues.

Fresh water organisms must be osmoregulators. If they were not, they would either lose all their ions to the ion-poor environment, or absorb so much water by osmosis, that they would swell and burst. Both of these outcomes would of course be disastrous and lead to certain death. Therefore, the evolutionary transition from sea water to fresh water could only have happened in organisms that are osmoregulators.

Organisms in terrestrial environments are faced with a different problem: the problem of conserving water. We already mentioned in Chapter 22 that terrestrial animals are at constant risk of losing water by evaporation from exposed surfaces. Loss of water leads to an increase in ion concentration of internal fluids. This is a significant problem for osmoregulators, but can also be a problem for osmoconformers, since there is a limit to the range of salinity levels such an organism can tolerate. A similar problem arises if water is conserved, but excess ions are ingested through food (e.g., if the food has a high salt content). Conversely, if the animal drinks too much fresh water, or if too much water accumulates via metabolic activity (we mentioned at

the beginning of this chapter that water is a breakdown product of many metabolic processes), ion concentrations can drop to below dangerous levels. Terrestrial organisms thus need to deal with a complex regulatory challenge: the need to conserve water—but not too much—and maintain a correct ion balance. All this regulation needs to be done while also dealing with removal of waste products.

The solution to this challenge in terrestrial animals is to excrete **urine**. Urine in terrestrial animals is a liquid or semi-liquid product that includes excess water, a carefully regulated combination of different excess ions and nitrogen waste in the form of urea or uric acid. Organisms in arid environments have evolved a suite of adaptations in their excretory systems allowing them to excrete waste products and excess ions without losing water at the same time. The waste products are either mixed with feces from the digestive system or deposited as a concentrated sludge. Note that the term "urine" is not reserved for terrestrial organisms and is used for the products of the excretory system in general as we will see below.

Osmoregulation can be even more complicated in some cases. Many mammals (including humans) regulate their body temperature by sweating. Sweat includes both salts and water and presents yet another factor that the excretory system has to respond to in regulating water and ion concentrations in the urine.

Types of excretory organs

The simplest excretory organs, found in several different phyla, are known as **protonephridia** (Figure 28.1A). A typical protonephridium includes several components. The internal part is usually a single-celled structure that sits within the internal fluid. This cell can either be a **flame cell** (named for its shape) or a **solenocyte**, with different organisms having one or the other. Flame cells have numerous cilia, whereas solenocytes have one or two flagella. The beating of the cilia or flagella creates decreased pressure in the cell, drawing water and small molecules into the cell. The liquid flows out of the cell into tubes that connect the cell to the outside via pores in the body wall known as nephridiopores. As the liquid flows down the tube, molecules and ions that need to be maintained are reabsorbed via specialized molecular pumps. Thus, the protonephridia remove fluid unselectively, while selectively returning some material. The fluid that is drawn into the excretory cell is known as **primary urine** and the final excretion product following reabsorption is known as **secondary urine**.

Metanephridia are multicellular structures found in a range of phyla (Figure 28.1B). The metanephridium starts with a ciliated funnel that opens into the organism's coelom or hemocoel. The cilia draw fluid from the coelom (primary urine) into the lumen of the metanephridium, which may be simple or coiled. Important ions and other molecules are removed from the fluid in a regulated manner and returned to the body. The excess material (secondary urine) is either excreted directly via nephridiopores or collected in a **bladder** and excreted via a urinary tract to a single opening.

Kidneys are complex vertebrate-specific organs. They are composed of numerous metanephridia (which in vertebrates are known as **nephrons**). Because vertebrates have a closed circulatory system, the fluid is not drawn from the coelom but from the circulatory system via a series of dedicated capillaries that reach individual nephrons, where blood is filtered into the nephron. Necessary components are reabsorbed and returned to the blood. The remaining waste products are collected from the nephrons through tubes that connect to a shared urinary duct, leading to the bladder. From there the secondary urine is removed from the body.

The excretory system of terrestrial arthropods is based on organs known as **Malpighian tubules** (Figure 28.1C). Intriguingly, very similar organs evolved convergently in insects and in myriapods. Indeed, the similarity in their structures was thought to be a key character uniting these two groups and indicating a shared terrestrialization event. This idea has now been discarded, given the phylogenetic distance between insects and myriapods and the evidence for independent terrestrialization of the two groups. Malpighian tubules are blind thin tubes that lie within the hemocoel. In all the previous examples, the excretory system collects fluid unselectively and removes necessary products and returns them to the body. In contrast, Malpighian tubules selectively remove

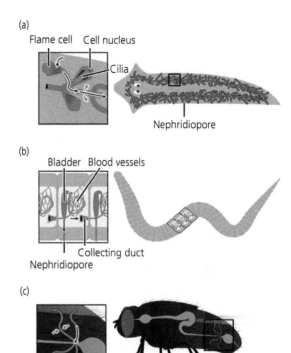

Figure 28.1 Different excretory organs: (a) Protonephridia, illustrated in a flatworm. (b) Metanephridia, illustrated in an earthworm. (c) Malpighian tubules, illustrated in a fly. The excretory system is indicated in green. Black arrows indicate flow of urine down tubes or vessels. Red arrows indicate movement of ions and waste products across vessel walls.

waste products and ions from the hemolymph in which they are immersed and carry them down the tubules. Uric acid crystals precipitate within the tubules and are carried with the other waste. The tubules connect to the digestive system at the midgut–hindgut border. The excretory waste products are transferred to the hindgut where they are mixed in with the excess material from the digestive system and removed as a semisolid mix of urine and feces.

The evolution of excretory systems

Excretory systems of some kind or other are found in all organisms. Unicellular eukaryotes have a rudimentary excretory system in the form of contractile vesicles that concentrate waste products and remove them from the cell via exocytosis. Non-bilaterian animals and members of Xenacoelomorpha (see Box 11.1) do not have separate excretory organs. However, most of them do have ammonia pumps that actively aid in removing ammonia.

The remaining bilaterians, to the exclusion of Xenacoelomorpha, are known as Nephrozoa, and have excretory organs as a unifying character. Protonephridia are found in most small animals and in animals without coeloms. There is some debate about whether protonephridia represent a more ancestral excretory system from which metanephridia evolved convergently several times, or metanephridia are the ancestral form and protonephridia evolved convergently by simplification

and miniaturization from metanephridia. The bulk of the evidence currently points to protonephridia being the original form, but this is not universally accepted. Vertebrate nephrons are probably homologous to metanephridia.

The advantage of a protonephridium is that its activity involves much less water loss than a metanephridium, and water loss is a more significant problem for small animals. Metanephridia are characteristic of larger coelomate organisms. As we saw above, terrestrialization led to the evolution of more complex excretory organs. Although these - excretory systems adapted for terrestrial living evolved independently in different lineages, they bear striking similarities, stemming from the shared requirements of a terrestrial excretory system.

CHAPTER 29

Vertebrate characteristics

An introduction to the vertebrates

Vertebrates make up only about 4% of total animal diversity in terms of species numbers. Nonetheless, they include some of the most familiar and best-loved animals. Vertebrates are bigger on average and attain larger maximum size than any other group of animals. This makes them dominant animals in almost any environment, if only for the sheer amount of food they need to consume and their environmental footprint. This is not meant to imply that an animal has to be large to have an effect on the environment. There are many examples of small animals with considerable impact because of their population size (e.g., ants and termites) or because they construct novel ecological niches (e.g., corals). Nonetheless, the significance of vertebrates far exceeds their low species numbers.

The invertebrate–vertebrate split is mostly a reflection of the increased interest devoted to vertebrates by humans. As important and as familiar as they may be, there is nothing fundamentally different between vertebrates and any other animal phylum. All of the general principles we have discussed apply to vertebrates as to any other phylum, and all of the organ systems we've covered are found in vertebrates, with their phylum-specific adaptations. However, vertebrates and invertebrates are usually taught in separate courses, discussed in separate textbooks (this book is an unusual exception) and often even studied in different academic departments.

Vertebrate innovations and body organization

The success of vertebrates is linked to a number of key innovations that set them apart from the invertebrate chordates (Figure 29.1). The most notable vertebrate novelty is **bone**. The term is used both for the tissue—a typical type of connective tissue strengthened with calcium phosphate—and for the structures formed by that tissue. Bone provides internal support and forms the vertebrate endoskeleton. This type of skeleton allows the attainments of large size, as internal skeletons are more efficient at larger sizes than external skeleton. The skeleton includes several components: the axial skeleton, the appendicular skeleton and the skull. The axial skeleton is based on the notochord (see Chapter 27), with the addition of repeated, interlocking units, known as **vertebrae**, which give the vertebrates their name. The appendicular skeleton includes elements of the limbs. The skull is built of a series of interlocking bones that enclose the brain and the main sensory organs.

From a developmental point of view, the key vertebrate innovation is the **neural crest**. This complex embryonic tissue type is derived from the margins of the forming hollow nerve cord or neural tube. The neural crest forms many crucial vertebrate tissues, including several types of bone, elements of the nervous system, pigment cells and others (see Chapter 31 for more details). In fact, the neural

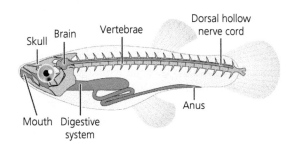

Figure 29.1 Vertebrate innovations, illustrated on a fish.

Organismic Animal Biology. Ariel D. Chipman, Oxford University Press. © Ariel D. Chipman (2024). DOI: 10.1093/oso/9780192893581.003.0029

crest contributes to so many structures, that some researchers say it should be seen as a fourth germ layer. **Sensory placodes** comprise another novel embryonic tissue type that is involved in the development of most vertebrate sensory organs. The combination of neural crest and placodes, together with an expanded anterior region of the neural tube—the brain—form the vertebrate head. The vertebrate head is a compound vertebrate novelty, which distinguishes the vertebrates from all other chordates. The suite of characters that characterize the vertebrate head have been called the "new head," to distinguish it from the simple anterior region of other chordates.

Like other chordates, vertebrates display **pharyngeal slits**. Ancestrally and in all extant primitively aquatic vertebrates, the **gill slits** (as they are more commonly called in vertebrates) are openings through which water passes from the pharynx to the gill as part of the gas exchange system. The gill slits are supported by cartilaginous structures known as **gill arches** (or **pharyngeal arches**). In terrestrial vertebrates (including secondarily aquatic vertebrates) the pharyngeal slits are transitory embryonic structures, but the gill arches remain as elements of the skeletal system. The gill arches change and move throughout evolution to give rise to jaws, jaw support structures, and additional anterior elements, including ear ossicles, laryngeal cartilage, and other structures.

The vertebrate body plan includes three separate segmental systems. The most obvious system is trunk segmentation, derived from embryonic units called **somites**. Trunk segmentation is true segmentation as we defined it in Chapter 18 and includes repeated elements of muscles, the skeletal system, the circulatory system, and the excretory systems (see more details in Chapter 31). Pharyngeal segmentation is less complex and is present only in the anterior part of the body. It includes the gill arches, gills, gill slits, and aortic arches (discussed in Chapter 30). The final segmental system includes only elements of the central nervous system. This is hindbrain segmentation (see below), which includes morphological and functional segmentation of the hindbrain, together with segmental nerves.

Vertebrate life history

As a rule, vertebrates reproduce sexually with rare exceptions. Separate males and females are almost universal in vertebrates. In some cases, there are only females. In even rarer cases (a few species of fish) individuals can change their sex during their lifetime, and a handful of fish species are simultaneous hermaphrodites. In those cases where there are only females, reproduction is through parthenogenesis. There are no vertebrates who reproduce asexually via budding or fission.

Regeneration of specific organs is found in some taxa. This is mostly regeneration of appendages or the tail. The famous case of lizard tail regeneration is not true regeneration, since the regenerated tail is not a full copy of the tail that was lost and does not include skeletal elements or normal integument. Whole body regeneration is not found in any species of vertebrates (unlike in other chordates), nor is regeneration of the head.

A biphasic life history with distinct larval and adult phases separated by metamorphosis is found in a few species of fish and famously in amphibians. These larvae are not the primary larvae we find in many marine invertebrates. They have a separate evolutionary history and are an innovation in each of the lineages where they are found.

Feeding and digestive systems

Vertebrates are ecologically diverse, and nearly all types of feeding can be found among them. Herbivory and predation are the most common type of feeding modes, with many macropredators and macroherbivores. Suspension feeding is found in many coral reef fish and in the largest animals on the planet, the baleen whales. Detritus feeding is found in many bottom-dwelling fish and rays. Scavenging is found both in fish and in a number of terrestrial mammals. Parasites are rare, and those that do exist are specialized blood feeders (e.g., vampire bats on the one hand and lampreys on the other) that don't form the long-term association with the host that is typical of parasites. There are no vertebrate endoparasites.

All vertebrates have a through gut with a mouth and an anus. The mouth almost always has jaws that usually include teeth—another vertebrate innovation. Jawless vertebrates either belong to the morphologically primitive Agnatha (see Chapter 30), or have secondarily lost them. Teeth are lost in all birds, and in some groups of fish, amphibians, reptiles and mammals. The stomach and intestine are distinct and usually separated by a muscular valve. As is typical for chordates, the anus lies at the base of the tail. Vertebrates normally have large digestive glands, including the liver and the pancreas. In addition, there are smaller glands that are taxonomically more variable. These include salivary glands in the mouth and various gastric glands in the stomach.

Some vertebrate digestive systems are highly complex and compartmentalized in order to deal with difficult to digest food items. These digestive systems often include a stomach with multiple compartments as well as an intestine that loops around the abdominal cavity to increase absorption area. Specific compartments in the stomach may act as fermentation centers, as in ruminant mammals. Muscular crops that help in preliminary break-down of food are found in many birds.

Circulatory systems

The vertebrate closed circulatory system is what allows them to achieve larger sizes than any other animals, as it allows distribution of nutrients and oxygen and removal of waste to and from all tissues, regardless of size. The circulatory system is based around efficient pumping by a muscular heart. Arteries lead from the heart to body tissues, and then divide to numerous efferent capillaries that reach almost every cell. Afferent capillaries merge to veins that lead blood back to the heart. In all vertebrates, the circulatory system is closely linked to the gas exchange organs, which may be either aquatic (gills) or aerial (lungs). In Chapter 30 we will discuss the evolution of the circulatory system and give more details on how it varies among different vertebrate groups.

Blood acts as an efficient oxygen carrier by using a dedicated oxygen-carrying protein known as hemoglobin. At the center of the protein is an iron-based organic molecule called heme. It is this iron-based molecule that gives vertebrate blood its red color. Hemoglobin is packaged in dedicated cells called red blood cells or **erythrocytes**. In most vertebrates, the red blood cells have a nucleus, but in mammals, they have lost the nucleus.

As in most animals, vertebrate blood also has a role in immune defense, with a series of different cell types involved in different aspects of the immune response. Because they lack hemoglobin, these cells are known as white blood cells. A range of hormones are dissolved in the blood, carrying different signals to target tissues throughout the body. These hormones control almost all organism-wide functions, including metabolic functions like blood sugar levels and heart rate, reproductive cycles, response to danger and stress, behavior, and other activities.

Finally, blood carries the breakdown products of the digestive system to provide nourishment to all the cells of the body. The blood also removes the metabolic products of the cells. All the blood from the whole body passes through the liver for detoxification of toxic compounds and for storage of excess energy sources. All the blood also passes through the kidneys for removal of excess ions and other waste and for balancing its water content.

Muscles, skeleton, and movement

All vertebrates have an internal skeleton. In most cases, this skeleton is composed of bone, although in some cases (e.g., lampreys, sharks and their relatives) it is composed of cartilage. The individual units of the skeleton, the bones, are connected to each other via joints, which are usually lined with cartilage. Joints can be simple hinge joints that allow movement in one plane or ball-and-socket joints, which allow more complex movements. Think of your elbow joint, a typical hinge joint, in comparison with your shoulder joint, a typical ball-and-socket joint. Movement of the joints is driven by striated muscles (also known as skeletal muscles—see Chapter 8). Most joints will have at least two muscles, and often many more, allowing a range of movement.

Vertebrate bones are divided into two types, based on their developmental origin. **Endochondral bones** are bones that start out as cartilaginous elements that gradually become ossified (converted into bone). The bone grows as it becomes ossified, until the animal reaches its maximum size. Most of the long bones in the vertebrate body are endochondral bones. This includes almost all of the appendicular skeleton, the vertebrae and others. **Dermal bone** forms within the mesodermal layer of the skin, the dermis (see Chapter 7). It usually ossifies directly from mesenchymal tissue, without passing through a cartilaginous stage. Most of the flat bones in the vertebrate body are dermal bones. This includes parts of the shoulder girdle and the pelvic girdle, most of the bones in the skull, the ventral portion of the ribs and others.

As a rule, vertebrates are mobile and active animals, with complex behaviors involving a very diverse range of movement. There are a few examples of sedentary vertebrates. These are mostly bottom dwelling fish that are ambush predators (predators that wait motionless for prey to arrive). Aside from these few exceptions, vertebrates tend to actively move in search of food and mates or to avoid danger. There are no cases of fully sessile vertebrates.

Most movement and locomotion in vertebrates is based on the motion of skeletal elements driven by skeletal muscle. There are also body wall muscles, but in most vertebrates, they have a relatively minor role in movement and locomotion. Most vertebrate locomotion is driven by the movement of appendages: limbs and fins.

Fins allow vertebrates to swim in a wide variety of modes depending of the size, shape, and position of the fins. Aquatic vertebrates range from highly maneuverable reef fish that can dart forward and backward quickly, through bottom-dwelling fish that glide just above the surface, up to fast-moving fish and aquatic mammals of the open ocean. In many fast-moving aquatic vertebrates the tail adopts a major role in locomotion, in which cases it becomes muscular and develops into an additional fin-shaped structure.

Limbs allow terrestrial vertebrates to walk, run, and jump with a range of specific locomotory patterns that can vary among closely related species.

Even within a single species, we can often see a range of locomotory patterns that are used for different environments and for different speeds (think of walking, trotting, and galloping in horses). Three groups of vertebrates independently evolved forelimbs capable of flight: birds, bats, and the extinct pterosaurs. The structure of the bones supporting the forelimb that has evolved into a wing in each of the three is completely different, testifying to their independent origin.

A small group of vertebrates primitively lack appendages—the agnathans (see Chapter 30)—and these swim by serpentine motion. Several groups of vertebrates have secondarily lost their appendages, and this is usually accompanied by an elongation of the body and a significant role for body musculature in locomotion. The best-known examples for this are snakes and eels, but there are many other similar cases.

Nervous systems

The main component of the vertebrate central nervous system is the dorsal hollow nerve cord that is typical of the chordates. The anterior part of the nerve cord is significantly expanded to give a hollow brain, which is responsible for many of the central and higher functions of the nervous system. The vertebrate brain is composed of three regions that differentiate during embryogenesis. Each of the regions is a separate expansion of the hollow nerve cord. They are known, from front to back, as the forebrain or **prosencephalon**, the midbrain or **mesencephalon**, and hindbrain or **rhombencephalon**.

There are of course significant differences in the relative size and importance of the three regions in various vertebrate taxa, and the roles of the regions are correspondingly different. What follows is a very general and simplified discussion of the structure and function of the brain, based mostly on the mammalian brain. The anterior part of the forebrain is called the **telencephalon** or **cerebrum**. It is an expanded region, with a pair of large fluid-filled spaces, or **ventricles**, surrounded by a layer of neural tissue, or cortex. In mammals, the cerebral cortex comprises most of the brain's volume, and it is where higher sensory processing takes place. The posterior part of the forebrain, or **diencephalon**, also expands to form a ventricle, albeit smaller than the cerebral ventricles. The regions surrounding the ventricles form the thalamus, hypothalamus, and a few additional regions, with roles in transferring sensory inputs between brain regions, hormonal control, circadian rhythms, and other activities.

The midbrain controls many sensory inputs and is responsible for much of the primary processing of data arriving from the sense organs. It also includes regions responsible for temperature control, sleep and wakefulness, and many other functions. Despite its varied functions, the midbrain in mammals is the smallest of the three embryonic brain regions, and does not include a fluid-filled vesicle.

The hindbrain is composed of two main regions: the **cerebellum** and the **medulla oblongata**. The cerebellum is a dorsal structure, which is responsible for balance, visual-motor coordination, etc. Its relative size is highly variable, forming most of the brain's volume in some fish and in birds, and conversely being a minor component in amphibians and some reptiles. The cerebellum is sometimes referred to as the small brain or lower brain. The medulla oblongata is the posterior most part of the brain and is sometimes referred to as the brain stem. It connects directly to the neural tube, with no clear boundary. The medulla is responsible for most autonomous functions: including breathing, heart rate, and blood pressure. The medulla is expanded to include a ventricle (the fourth brain ventricle), with a very thin roof. During embryonic development, the hindbrain is transiently divided into segmental structures known as **rhombomeres**. Although they are morphologically indistinct in the adult brain, they are linked with a series of cranial nerves that emanate from the hindbrain, and can be seen as a distinct, if minor, segmental system.

Although the vertebrate central nervous system is not inherently segmental (unlike what we saw in annelids and arthropods), it does have segmental elements. These are a series of ganglia outside the nerve cord and lateral to it on both sides, known as the **dorsal root ganglia**. These ganglia are derived from the embryonic neural crest and are mostly composed of sensory neurons. In addition to the segmental dorsal root ganglia, elements of the autonomous nervous system, specifically the so-called sympathetic nervous system, are also arranged segmentally within the nerve cord.

Vertebrates have some of the most complex sense organs known from animals. The sense organs in the head: eyes, ears, and nose, are derived from sensory placodes and form a part of the vertebrate "new head" (see above). The vertebrate eye is an image-forming eye that includes a lens that can change shape through muscle activity to vary the focal length and focus on objects at different distances. It has a pupil that can dilate and contract to allow more or less light through in response to varying light conditions. It has a retina composed of thousands to millions or photoreceptive cells, each with its own axon leading to processing centers in the brain. Different vertebrates have varying combinations of types of photoreceptor cells allowing vision at different wavelengths and light intensities, and in many cases even forming color images.

Vertebrate ears are originally vestibular organs, responsible for relaying information about the organism's position in space and its movement and acceleration. In terrestrial vertebrates the vestibular organs adopt a secondary and no less important role in sound reception and interpretation. In some primitively marine vertebrates, the ear already has a minor role in sound reception, but this is much expanded in the transition to terrestrial habitats. We've already discussed the structure and function of the marvelous vertebrate cochlea in Chapter 12. The acuteness of hearing in different vertebrate taxa depends of the exact structure of the cochlea, on the size and shape of the ear drum, which is the organ that first picks up air vibrations, and in the number and shape of the ear ossicles—a series of small bones that transmit the vibrations picked up by the ear drum to the cochlea. Mammals have three ossicles, derived from primitive gill arches, and all other terrestrial vertebrates have one. The size and shape of that one ossicle (the stapes) can be very different, leading to different sensitivities.

We also discussed the function of the vertebrate olfactory organ, the nose, in Chapter 12. In aquatic vertebrates, the naval cavity opens only to the surrounding water, and the nasal epithelia pick up chemical signals from the environment. In most terrestrial vertebrates, the nasal cavity is connected to the pharynx via an opening known as the **choana**. This allows terrestrial vertebrates to breathe through the nose, and also allows the nasal epithelia to pick up chemical signals from the food being processed in the pharynx. Chemosensory receptors are also found on the tongue itself, and these mostly pick up dissolved chemicals. Some fish have chemosensory receptors on the integument, but for most vertebrates the nose and tongue provide the only chemical sensation.

Beyond the cranial sensory organs, vertebrates have sensory organs distributed over the entire body as touch receptors and pain receptors on the integument, and proprioceptors in the joints. There are pain receptors in internal organs as well, but these are poorly understood. The lateral line system is a distributed mechanosensory system found in primitively aquatic vertebrates. More details on this system were also given in Chapter 12.

CHAPTER 30

Vertebrate diversity

Introduction to vertebrate diversity

There are between 60,000 and 70,000 species of vertebrates (Figure 30.1). About half of these are fish and half are terrestrial vertebrates (tetrapods). Vertebrates are traditionally divided into six to seven classes, but most of these classes are paraphyletic, and this traditional division is not a good reflection of vertebrate diversity. Rather than following this traditional classification, we will trace the diversity of vertebrates by following their evolutionary history, pointing out important transitions and innovations along the way. Because of their size and because of the increased interest science has had in these animals, our knowledge of vertebrate diversity is very good, and the proportion of undescribed species is probably much lower than in any other phylum. As a result of their easily fossilizable bony skeleton, vertebrates have a good fossil record, and therefore our understanding of their history and the relationships between the different taxa is also better than in most animal groups.

Fish

Like all other animals, the vertebrate story begins in the sea. Among the earliest fossil chordates represented in the Cambrian fossil record, there are already species with a clear expansion of the anterior neural tube, suggesting they are part of the lineage leading to vertebrates. By the Ordovician we see the first appearance of the key defining feature of the vertebrates: bone tissue. Surprisingly, bone first appears not as an internal skeleton, but as external armor. Throughout the Ordovician fossil record, we find a diversity of armored fish representing a range of ecological specializations and feeding

habits. A common hypothesis about the origin of bone posits that it first appeared as a mineral reservoir for calcium and phosphate. Since it was stored in the dermis, it also provided protection, and was thus elaborated through natural selection to give the massive armor we see in many of the early vertebrates.

Notably, these early armored fish had no jaws and no fins. A small group of extant vertebrates still preserves the ancient ancestral vertebrate body plan with no jaws and no fins. These are classified as Cyclostomata, the jawless fish, including today only lampreys and hagfish. This group together comprises just over 100 species.

Paired fins first appeared in a lineage of Ordovician armored fish. The pectoral (anterior) fins appeared first, followed rapidly by pelvic (posterior) fins. Somewhere within this lineage of fish there appeared the most significant vertebrate innovation, which probably led to their becoming a dominant marine group: the **jaws**. Jaws evolved through the modification of the first gill arch of their ancestors. The second gill arch was modified to give a support structure that connects the jaws to the skull. The Ordovician fish with jaws and fins are the ancestors of all jawed vertebrates found today. Jawless fish continued to exist alongside jawed fish, but became increasingly less diverse as they were outcompeted by the jawed fish, or Gnathostomata.

As fish became more active, the heavy external armor was gradually reduced and replaced by the internal skeleton we recognize as a vertebrate character today. By the Devonian, many different lineages of vertebrates had diversified in the sea, leading to the Devonian sometimes being referred to as "the age of fishes." By this time, we can identify three major lineages that lead to some of

Organismic Animal Biology. Ariel D. Chipman, Oxford University Press. © Ariel D. Chipman (2024). DOI: 10.1093/oso/9780192893581.003.0030

Figure 30.1 Some examples of vertebrate diversity: (a) A teleost fish, the emperor angel fish *Pomacanthus imperator*. (b) A lissamphibian, the European pond frog *Rana temporaria*. (c) A lepidosaur amniote, the sand gecko *Stenodactyllus petrii*. (d) an archosaur amniote, the white pelican *Pelecanus onocrotalus*.

Source: a: Photo supplied by Shutterstock: Victor1153 (https://www.shutterstock.com/image-photo/fish-angel-imperial-emperor-angelfish-on-1593866962); b: Photo supplied by Shutterstock: Addictive Creative (https://www.shutterstock.com/image-photo/european-common-frog-rana-temporaria-standing-2232915465); c: Photo supplied by Shutterstock: Kurit Afshen (https://www.shutterstock.com/image-photo/sand-geckostenodactylus-petrii-sunbathing-basking-2135229533); d: Photo supplied by Shutterstock: meunierd (https://www.shutterstock.com/image-photo/great-white-eastern-pelican-rosy-bird-688823338).

the extant vertebrate classes. Cartilaginous fish or Chondrichthyes split from bony fish very early in vertebrate history. The ancestors of today's cartilaginous fish had bony skeletons. They are derived (like all vertebrates) from armored Ordovician fish, and their history includes a stage with a bony internal skeleton. Chondrichthyes is represented today by sharks, rays, and chimeras (rabbit fish). Although the lack of bone in their skeleton is often represented as being a primitive character, and sharks are often described as "ancient," placing them in an evolutionary context shows that the cartilaginous skeleton is an innovation within this group. Sharks are of course no more ancient or young than any other vertebrate group.

After the branching off of the ancestors of sharks and rays, there was a second major split within vertebrates. One branch of this split, the Actinopterygii or ray-finned fish, led to most of the animals we refer to as fish today. The second branch, the Sarcopterygii or lobe-finned fish, led to a small group of "fish" and to terrestrial vertebrates. The distinction between these branches lies in the structure of the fins. The ray-finned fish have no bone in their fins, the fin skeleton being instead composed of keratinous rays. They also have no muscles within the fins. All the movement of the fins is driven by muscles that lie in the body wall, external to the fins. In contrast, lobe-finned fish have fins with bony skeletons and these bones are connected and activated by muscles, giving the fins a fatter lobe shape.

There are over 30,000 species of ray-finned fish, nearly half of all vertebrate diversity. They are found in the sea and in fresh water. They have diversified into numerous shapes and sizes and are found in all aquatic habitats, with the exception of some ephemeral water bodies. The largest group within ray-finned fish is teleost fish, characterized by the evolution of a novel jaw-hinge mechanism that allows them to open their mouths very widely, creating a region of lower pressure that draws food into the mouth.

In the Devonian, the diversity of lobe-finned fish rivaled that of ray-finned fish. However, there are only a handful of lobe-finned fish living today, including lungfishes and the enigmatic coelacanths. However, the lobe-finned lineage is important in vertebrate history, since within this lineage are the ancestors of all terrestrial vertebrates, or Tetrapoda.

The group that is colloquially referred to as "fish" is a paraphyletic assemblage unified only by the fact that their ancestors were never terrestrial. Fish are composed of three monophyletic groups: jawless fish (Cyclostomata), cartilaginous fish (Chondrichthyes), and ray-finned fish (Actinopterygii), and one paraphyletic group: the lobe-finned fish.

Vertebrate terrestrialization

The origin of Tetrapoda lies within Devonian lobe-finned fish. They arose from large fresh-water predators, probably similar in habits to crocodiles of today. The transition to terrestrial life took place over a few tens of millions of years in the late Devonian. This transition is very well documented in a series of fossils, many discovered only in the last few decades, that show a gradual transition of several different skeletal characters from typical "fish-like" structures to elements adapted for life on land. Surprisingly, many of these transitions occurred while the early tetrapod relatives were still fully aquatic. As they lived in inland water with low oxygen content, they evolved rudimentary lungs before they made the transition to land (a situation still found in modern lungfish), and because they lived in shallow water with poor visibility, they

evolved limbs that were able to push them along the bottom of pond and support their weight enough to lift their head above the water. The transition to life on land occurred in animals that already had four strong walking limbs (tetrapods literally means "four legs") and full air breathing.

A number of skeletal modifications took place in early tetrapods and their ancestors. The most important of these was the transition of the lobe-fin to a limb. This involved an elongation of the proximo-distal axis and its subdivision into distinct regions that would ultimately become the series of bones that make up the typical tetrapod limb. In parallel was the evolution of pelvic and pectoral girdles that became connected to the vertebral column and provided support for the new limbs. With the formation of the pectoral girdle, the head that was fused to the vertebral column separated, with the evolution of a neck that allowed the head to move freely. Changes to soft tissue and to physiology are not normally preserved in the fossil record, but we can assume that concomitantly with these skeletal changes, a modified circulatory system evolved, transferring the main site of blood oxygenation from the gills to the lungs.

By the early Carboniferous, there was a diversity of terrestrial vertebrates living in different habitats, but all still closely tied to the freshwater habitats in which they first evolved. These Carboniferous tetrapods are usually referred to as amphibians, but we must be careful with this term and clarify its meaning. Amphibians are animals that divide their time between land and water, normally returning to water to reproduce and for the first stages of their life history. However, these Carboniferous amphibians were not the animals we call amphibians today (we discuss those in the next paragraph). They were medium- to large-sized animals, reaching maximum sizes between a few tens of centimeters and a couple of meters. They were almost exclusively predators, feeding on arthropods and fish. Although at least some of them went through a biphasic life cycle with an aquatic larval stage and a terrestrial adult stage, they include the ancestors of all extant terrestrial vertebrates, not just of modern amphibians. Thus, if we use the term "amphibian" to include both Carboniferous and modern taxa, we must acknowledge that it is a paraphyletic group.

Extant amphibians (technically known as Lissamphibia) comprise three taxa: Anura—frogs, toads, and their relatives; Urodela (or Caudata)—salamanders and newts; and Gymnophiona (or Apoda)—the limbless caecilians. There are over 8,500 described species of amphibians, many of them discovered only in the last few decades in tropical forests worldwide. Most amphibians have a thin integument that allows water loss, so they must live in damp habitats or close to water. Their thin integument also makes them susceptible to environmental toxins, and they are therefore among the most vulnerable vertebrates and the most prone to extinction following habitat degradation. Most amphibians have distinct tadpole and adult stages, but several lineages have evolved direct development that skips over the tadpole stage.

Amniotes

By the late Carboniferous, drying conditions worldwide promoted the evolution of better adaptations to life on land. These adaptations included modifications to the integument and the evolution of terrestrial reproduction. Changes to the integument involved the elaboration of the keratinous outer layer of the epidermis to form a water-proof covering. In many cases, this took the form of distinct ectodermal scales, sometimes coupled with dermal scales. Thus, hundreds of millions of years after losing external armor, some vertebrates re-evolved it.

While modifications to the integument made it possible to live on land, the more important evolutionary innovation was the appearance of the **amnion**, a reproductive structure that lends its name to Amniota—the fully terrestrial vertebrates. The amnion is a liquid-filled and water-proof sac within the egg that allows the embryo to develop in an aquatic environment even when the egg is laid outside of the water. The fossil record does not provide any information about how this structure evolved and what intermediate stages it went through. Nonetheless, there are a number of skeletal characteristics that we can associate with the group within which the amnion evolved, and which allow us to trace the evolutionary history of the amniotes, even without direct evidence of the amniotic egg itself. Further elaboration of the amniotic

egg included a tough, desiccation-resistant outer shell, and an internal waste-storage compartment known as the **allantois**.

The amniotes evolved to be much more active than their ancestors, and their skeleton evolved to be lighter in weight and more mobile. One aspect of the reduction of the weight of the skeleton was the appearance of openings in the skull, known as **skull fenestrations** (from fenestra, the Latin word for window). One major lineage evolved two skull fenestrations, one above the other, in the posterior of the skull. This lineage is known as Diapsida. Another major lineage evolved one posterior ventral opening, and it is known as Synapsida. Each of these two lineages gave rise to a major branch of vertebrate diversity. The early amniotes with no skull fenestrations are known as Anapsida—a paraphyletic group.

Dinosaurs, birds, and "reptiles"

The diapsid lineage in turn split into two main lineages—Archosauria and Lepidosauria. The first of these lineages, the archosaurs, rose to prominence in the Triassic, with a number of different archosaur taxa replacing each other in the earliest part of the Triassic. However, one group emerged from the early Triassic as clearly dominant—the dinosaurs. Dinosaurs are probably the most famous fossil organisms. Their large size coupled with the treatment they have received in popular media has made them icons of paleontology. However, beyond their size, there is nothing inherently different about dinosaurs compared with any other fossil taxon, not in the number of species known nor in their general morphological diversity or ecological specializations. It is mostly their importance in public interest in science that makes them stand out.

We will not discuss the diversity of dinosaurs further, but will focus on one specific group. Theropoda was the main lineage of meat-eating dinosaurs. Early dinosaurs were all predators, but all other lineages aside from the theropods transitioned to eating plants. Mega-carnivores like *Tyrannosaurus* and *Allosaurus* are the most famous theropods, but there were also many smaller members of this group. Some lineages of small theropods became very active with elevated body temperatures and therefore evolved a form of insulation to maintain the elevated temperature. This insulation started out as fibrous down but gradually evolved into **feathers**. The structure of feathers allows them to trap air more efficiently, and thus to act as better insulators, and also provided an excellent starting point for the evolution of aerodynamic surfaces that allowed small theropods with complex feathers to glide and ultimately to fly. These feathered theropods are the ancestors of the diverse group we know as birds or <u>Aves</u>. The transition from small running predators to flying animals is well documented in the fossil record. Birds were already flying and fairly diverse by the late Cretaceous. The mass extinction event at the end of the Cretaceous (see Box 30.1) led to the extinction of almost all dinosaurs, with the exception of one group of small flying theropods—the birds.

After the extinction of the dinosaurs, birds underwent a significant evolutionary radiation. Today, they are the largest or second largest extant class of terrestrial vertebrates with over 10,000 species. Birds are characterized by modified forelimbs that function as wings and by feathers. They have adapted to many different modes of flight from hovering to flapping to soaring, mediated by different size and shape of the wings. Some birds are flightless and have reduced or vestigial wings. In addition to the modified forelimbs, they also have characteristic hindlimbs with four fingers ending in claws. Birds have lost their teeth and instead evolved a keratinous beak. The size and shape of the beak are closely linked to the food source and feeding mode, from thick beaks for cracking seeds, to thin beaks for catching insects and worms and up to the large sharp beaks of raptors used to rip flesh.

Another group of archosaurs that achieved brief dominance in the early Triassic includes the ancestors of extant crocodiles and alligators. Despite relatively low diversity, this group survived the end-Cretaceous mass extinction. There are less than 30 species of crocodiles and alligators today. All have amphibious lifestyles (but they are not amphibians!) and all are predators. They have specialized as ambush predators in shallow water with the different species having very similar body plans and ecological niches in different parts of the world.

The turtle lineage includes about 360 species. Some are fully terrestrial, some are amphibious or freshwater adapted, and some are marine. There is a debate over whether turtles originally evolved as terrestrial animals or as aquatic animals, and different fossils tell different stories about their ecological origin. Fossils of early turtles suggest they emerged within the diapsid branch (despite having a skull with no fenestration), probably closer to the archosaurs.

The last group of archosaurs worth mentioning is the pterosaurs, the first vertebrates to fly. They lived alongside the dinosaurs, diversifying in the late Triassic at about the same time as the dinosaurs, and went extinct in the late Cretaceous, shortly before the dinosaurs. They are thus often lumped with dinosaurs in popular opinion, but are in fact a distinct taxon. Some pterosaurs were as small as sparrows and probably occupied similar ecological niches. Other pterosaurs reached wingspans of several meters, were as tall as giraffes, and were the largest animals ever to fly.

We move now to the second diapsid branch, Lepidosauria. While lepidosaurs lived alongside the dinosaurs and other archosaurs, they never reached the same sizes or the level of diversity as dinosaurs. It was only after the extinction of the dinosaurs that they became prominent. Lepidosaurs today include lizards and snakes, as well as the lizard-like tuatara of New Zealand (a sole representative of a once much larger group). Most are fairly small, with the only exceptions being the giant constrictor snakes such as boas and pythons, and some species of varanid lizards like the Komodo dragon. There are several known cases of limb loss within the lepidosaur lineage, the most successful being the one that gave rise to snakes. Indeed, there are several Cretaceous fossils of snakes with reduced limbs, shedding some light on the evolution of snakes.

All of the animals we've discussed in this section are known generally as reptiles, with the exception of birds. As we can see, reptiles are a diverse assemblage with different evolutionary histories and relatively little in common. Reptiles are also a prime example of a paraphyletic group, because the birds are members of this group, but usually not counted within it.

We can't end our discussion of reptiles without mentioning marine reptiles. Marine reptiles are amniotes that secondarily returned to the marine environment. This happened many times (probably about 10 times) in different lineages at different points in amniote history. During the Jurassic and Cretaceous, while dinosaurs were the main terrestrial vertebrates, there were a number of lineages of large marine reptiles, including ichthyosaurs, plesiosaurs, nothosaurs, and mosasaurs, to name a few. The only remaining marine reptiles today are the turtles, and a few of species of sea snakes.

Mammal evolution

Let us now return to the main amniote split and discuss the second major lineage, the synapsids, which is the lineage from which mammals arose. The evolutionary history of the synapsids starts before the dinosaurs. They were the dominant land animals of the Permian, undergoing a series of sequential evolutionary radiations. Among the Permian synapsids are the sail-backed reptiles such as *Dimetrodon* and a diverse series of animals popularly known as "mammal-like reptiles." Throughout this period, the fossil record indicates a gradual reduction in jaw bones and a transition in the structure of the jaw. The elements in the lower jaw were reduced so that only the large dermal bone known as the dentary remained, and many of the original jaw elements shifted posteriorly to initially give rise to a novel jaw hinge. Later in synapsid evolutionary history, this hinge was replaced by a hinge between the dentary and the skull, and some of the original jaw bones were reduced to give rise to the ear ossicles (see Chapter 29). Interestingly, the evolutionary transition from jaw bone to ear ossicle is also mirrored in mammalian embryonic development.

During the early Triassic, the synapsids were gradually replaced by archosaurs as the dominant terrestrial vertebrates. In the Jurassic and Cretaceous archosaurs (mainly dinosaurs) became large, while synapsids became small and nocturnal. During this time, they evolved acute vision, and relatively larger brains to deal with nocturnal living. As they became smaller, they also had to evolve insulating structures to maintain elevated body temperatures (just like the birds did). Hair and fur probably first appeared as sensory structures, but

were elaborated and adapted for thermoregulation. It is during this period that synapsids gradually evolved the suite of traits we recognize as mammalian. Behavioral traits like maternal care probably evolved at about the same time, though there is of course no direct evidence for this in the fossil record. These early mammals still laid eggs, but probably already had specialized ventral glands for feeding their young—the precursors of mammary glands. Fossil discoveries from recent years show that despite the dominance of dinosaurs, mammals remained diverse, with different lineages adapting to life in trees, in burrows, and even in semiaquatic environments.

The extinction of the non-avian dinosaurs opened numerous ecological niches, which the mammals (or Mammalia) rapidly evolved and diversified to occupy. Within a few million years, they had diversified to fill a wide range of ecological specializations, in parallel with growing significantly in size. Extant mammals are divided into three lineages that were already distinct by the Cretaceous. The earliest branching and smallest of these lineages is Monotremata—platypuses and echidnas. Monotremes are confined to Australia and nearby islands and include only five known species. They preserve a number of ancestral characters from early mammals, such as egg laying and ventral lactation. However, it is important to note that although some monotreme characteristics can be seen as "primitive," in other aspects they have undergone significant evolution and display many unique and specific adaptations.

Marsupialia, the second mammalian lineage, evolved in what is now South America and Australia. In Australia, marsupials and monotremes remained the only mammals and marsupials thus evolved to fill all suitable niches. Marsupials do not lay eggs, and the embryo develops within a uterus. The defining character of marsupials is the presence of a maternal pouch or **marsupium**. Offspring are born very precociously and crawl to the pouch where they cling to a **mammary gland** or **teat** for nourishment. They feed and grow until they are large enough to emerge from the marsupium. In most cases, the offspring continue to return to the marsupium for an extended period of maternal care.

Placental mammals (<u>Placentalia</u> or <u>Eutheria</u>) are the third and largest lineage, and they make up the majority of mammals found today. Placentals are characterized by a long pregnancy, with the embryo being connected to the mother's circulatory system via a shared organ known as the **placenta**. The embryo derives nourishment through the placenta, allowing it to develop to a much more mature stage than marsupials and many other vertebrates. The placental mammals underwent a series of localized adaptive radiations in different parts of the world following the end-Cretaceous extinction. Each of these radiations led to one of the four main lineages of placental mammals. <u>Afrotheria</u> radiated in Africa and gave rise to elephants, hyraxes, sea cows, and a number of small insectivorous taxa. <u>Xenarthra</u> radiated in South America and gave rise to armadillos, sloths, and anteaters. <u>Laurassiatheria</u> radiated in the northern continent that ultimately split to form North America and Eurasia. It includes hoofed mammals (ungulates), the mammalian carnivores, bats, and several groups of insectivores. <u>Archontoglires</u> probably also diversified in the northern part of the planet, but this is not clear. It includes rodents and their kin, and primates, including ourselves.

The evolution of breathing and circulatory systems

One of the systems that underwent the most significant series of transformations throughout vertebrate evolution is the gas exchange and circulatory system. Having followed the evolution of vertebrate diversity, we can now trace how this system evolved with the changing modes of life in the different lineages.

The ancestral vertebrate heart is an aquatic heart, adapted to gas exchange through anterior gills. It is composed of two regions: a muscular **ventricle** and a thin-walled **atrium** lying dorsal-posterior to it. The ventricle contracts rhythmically and drives blood anteriorly through a main artery, the **ventral aorta**. The aorta splits to deliver blood to the individual gills where it is oxygenated and releases carbon dioxide. In jawed vertebrates, there are five pairs of gills and thus five pairs of arteries, sometimes known as **aortic arches**, leading to the gills.

The vessels carrying oxygen-rich blood back from the gills reconnect at the **dorsal aorta**, which heads posteriorly along the body to supply blood to all the tissues. Oxygen-poor blood re-collects through a series of veins that pass through the liver and the kidneys before reaching the heart, entering it via a pair of veins that connect to the atrium. Blood is drawn from the atrium into the ventricle as the ventricle expands, and the cycle begins again.

The circulatory system of extant lungfish represents an example of the transitional state probably found in early tetrapod ancestors. In lungfish, one of the aortic arches leads to the lung, instead of to a pair of gills. Blood that is oxygenated in the lung returns to the heart, and enters it through the atrium, which is separated into two parts. The blood returning from the lung enters from the left part of the atrium. The ventricle is also partially separated into two parts, so that the oxygenated blood is not pushed toward the gills, but to a separate artery that leads directly to the dorsal aorta where it mixes with blood oxygenated in the gills. The blood circulates through the body as described above and returns to the heart via the right part of the atrium.

Modern amphibians do not have gills as adults. They therefore lose most aortic arches following metamorphosis. The arch that leads blood to the lungs (now paired, compared to the single lung of lungfish) becomes the main aortic arch and is known as the pulmonary artery. Oxygenated blood returns to the heart, where there are distinct left and right atria, via the pulmonary vein. The pulmonary vein enters the left atrium. Deoxygenated blood from the body enters the heart via the right atrium. As in lungfish, there is partial separation of the ventricle into a left and right region. Oxygenated blood in the left region is pumped directly to the dorsal aorta, whereas blood in the right region is pumped toward the lungs.

The original situation in amniotes is basically similar, with an elaboration of the separation between the right and left halves of the ventricle and more distinct left and right atria. The separation between oxygenated and non-oxygenated blood is almost complete, despite the lack of separation between the left and right halves of the ventricle. This is achieved through the flow patterns of the blood that drive oxygenated blood into the aorta and deoxygenated blood into the pulmonary artery. This is the situation seen in modern lepidosaurs and in turtles.

In modern archosaurs (crocodiles and birds) and in mammals there is full separation between oxygenated and non-oxygenated blood with two distinct ventricles, each connecting to a single aorta. In essence, there are two separate circulatory systems: a pulmonary system that leads blood to the lungs and back and a systemic circulatory system that leads blood to the rest of the body and back. Despite the similarity between the archosaur and mammalian systems, they probably evolved convergently from the ancestral amniote system.

Box 30.1 The end-Cretaceous mass extinction

As discussed in Box 26.1, mass extinctions events are important factors in the evolution of life. The end-Permian mass extinction marks the border between the Paleozoic and the Mesozoic. The extinction at the end of the Cretaceous period, 65 million years ago, marks the border between the Mesozoic and the Cenozoic eras. Both extinctions led to a change in the makeup of the biosphere that was significant enough that we see the world before and after the extinction as belonging to different geological eras.

The end-Cretaceous mass extinction, also known as the K/T extinction, is the most recent generally accepted mass extinction event. It is also the best understood. Unlike most other events, which took place over an extended period and were driven by compound factors, the evidence suggests that the K/T event was the result of a singular cataclysmic event, caused by the impact of an extraterrestrial object. The impact occurred close to what is now the Yucatan peninsula in Mexico. The force of the impact led to massive shock waves and tsunamis, and ejected large amounts of steam, ash, and molten rock into the atmosphere. Light from the sun was blocked or dimmed for several years, leading to a reduction of photosynthetic activity and a collapse of global food chains. Those animals that were not killed by the immediate aftermath of the impact, suffered and died in the ensuing disaster. It is not clear whether the environmental crisis lasted only a few years until the atmosphere cleared, or whether it continued for hundreds or thousands of years afterward.

Although this extinction event was rapid and dramatic, it was also relatively brief, and recovery was much swifter than the recovery from the end Permian event. Some 75% of all animal species were driven to extinction, but looking at higher taxonomic levels, there are relatively few animal taxa that disappeared completely. The most famous group of animals to go extinct is of course the dinosaurs, but recall that in fact, the entire group did not disappear, since birds are direct descendants of the dinosaurs.

In the sea, the most notable group to go extinct is the ammonites, a group of shelled cephalopods. Many groups of marine reptiles also disappeared, but many of them were already in decline before the impact, and some may had already gone extinct before the impact.

In the aftermath of the extinction, mammals radiated quickly and within a few million years had filled all of the ecological niches vacated by the large dinosaurs. Within 10 million years more, one group of mammals, the whales and their kin, had returned to the sea and filled niches vacated by marine reptiles. Birds had already evolved before the end of the Cretaceous, but they too diversified significantly in the aftermath of the extinction. The diversity of terrestrial invertebrates did not change quite as dramatically. However, from the vertebrate perspective, the mammal and bird dominated world we live in was largely shaped by the K/T event.

CHAPTER 31

Vertebrate organogenesis

Understanding the vertebrate body plan

We complete our discussion of the vertebrates with a chapter devoted to the vertebrate body plan. To understand the vertebrate body plan we will take a different approach to that we have taken with other phyla. We will follow the development of the different organs in the vertebrate body in order to understand their structure and their spatial arrangement. This chapter repeats some of the discussion in Chapter 25, but goes into more detail and describes the specifics of vertebrate development, as opposed to the previous generalized description. The processes described in this chapter also serves as a way to tie together different ideas that we have covered throughout this book.

The frog embryo as a model

For our description of vertebrate organogenesis, we will use the frog embryo as a model. We will follow it from the blastula stage (see Chapter 25 as a reminder of the different terms used in this chapter) up to the tailbud stage. Frogs have relatively conservative development and are thus a good example of a "typical" vertebrate. The frog egg is spherical making development easy to follow and understand. We will highlight specific points where the development of other vertebrates may differ from that of the frog.

If we were to follow the development of other vertebrates, we would see essentially the same process. The relative size and precise topological relationships are different among different vertebrate classes. In most amniotes, the egg is large and filled with yolk, leading to a flattened embryo and a modified developmental process. Mammalian early development is highly divergent relative to other vertebrates, although late stages of organogenesis are roughly similar. A well-studied phenomenon is that despite these differences in early development, all vertebrates converge on a broadly similar stage in mid-development, the so-called **phylotypic stage**. Development before this stage varies depending on the size and shape of the egg or embryo. Development after this stage diverges to give rise to taxon-specific differences. However, during the phylotypic stage, the general and conserved aspects of the vertebrate body plan are established.

Gastrulation and germ layers

We will start our description with the blastula. A frog blastula is a ball of cells with an internal cavity, the blastocoel. Depending on the species, it ranges in size from just under a millimeter to a few millimeters in diameter. Gastrulation begins with the movement of cells from the outside of the blastula into the blastocoel cavity. This movement, or involution, begins at a single point in the dorsal midline of the blastula, the blastopore. As cells involute they undergo a transition and give rise to the mesoderm. Involuting cells push the blastocoel cavity, which shrinks and ultimately disappears, while creating a new cavity, the **archenteron**. The archenteron develops as an elongated tube that is the basis of the digestive system.

Involution begins only on the dorsal side of the blastopore, but as gastrulation progresses, it continues on all sides of a roughly circular blastopore opening. If you recall our earlier discussion on embryogenesis (Chapter 25), you will remember

Organismic Animal Biology. Ariel D. Chipman, Oxford University Press. © Ariel D. Chipman (2024). DOI: 10.1093/oso/9780192893581.003.0031

that the blastopore will ultimately form the posterior end of the digestive system, the anus, since vertebrates are deuterostomes. Cells that involute at different stages of the gastrulation process will end up at different locations. The earliest involuting cells will contribute to the head mesoderm, while late involuting cells will contribute to the mesoderm of more posterior regions.

By the end of gastrulation, all cells in the embryo have undergone preliminary differentiation to the three germ layers. As mentioned above, most of the involuting cells differentiate into early mesoderm. Cells on the ventral side of the archenteron close over it to give the early digestive tract and differentiate into endoderm. Cells that did not involute during gastrulation, but remained outside on the embryo's surface, form the basis of the ectoderm. The precise mode by which cells move during gastrulation and the way they enter the blastocoel varies among vertebrate classes, but the determination of the germ layers during gastrulation is conserved.

The dorsal lip of the blastopore, the region where cells first start the gastrulation process, has a special role in development. It has been known for over a century, based on landmark experiments carried out by Hans Spemann and his student Hilde Mangold, that this region acts as an **organizer** that influences the formation of the embryonic axis. More recent work using molecular techniques has been able to identify the precise components that are involved in the activity of the organizer.

The neural tube and neural crest

At the end of gastrulation, the dorsal side of the gastrula flattens and thickens to give rise to the **neural plate**. In parallel, the embryo starts to elongate, so it is no longer spherical but more ovoid. The elongated embryo is now usually referred to as a **neurula** (Figure 31.1). The blastopore, at which gastrulation began, is pushed backward to form the posterior end of the embryo.

The neural plate then starts to fold, with the lateral margins rising and ultimately closing together. The elongated and closed neural plate now becomes the neural tube. The neural tube sinks and is covered

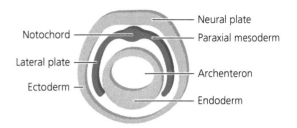

Figure 31.1 Cross section through an early neurula-stage frog embryo: Blue—ectoderm, red—mesoderm, yellow—endoderm.

by ectoderm from both sides. This structure is the embryonic basis of the hollow dorsal nerve cord.

As the neural plate folds, cells from its most lateral margins end up in the most dorsal position, because the plate folds into a tube (Figure 31.2, Figure 31.3). These dorsal cells start to separate and migrate away from the forming neural tube. These cells will form the neural crest. Neural crest cells gather into clusters and migrate in a series of streams along anterior posterior axis of the embryo. They reach specific positions in the embryo and start differentiating into a wide variety of different tissues. They form the cartilaginous gill arches in the head region and other components of the face. A subset of anterior neural crest cells forms the cells that drive the differentiation of teeth. Trunk neural crest cells differentiate into the precursors of melanocytes or pigment cells. Other populations of trunk neural crest cells give rise to the segmentally arranged dorsal root ganglia and sympathetic ganglia (see Chapter 30). Non-segmental neural crest cells contribute to the parasympathetic ganglia and to various ganglia and glands of the digestive and endocrine systems.

Somites and the segmented vertebrate body

At the same time that the neural plate and neural tube differentiate, the mesoderm also differentiates into several parts (Figure 31.1, Figure 31.2). The **axial mesoderm** differentiates in the medial-dorsal part of the embryo to give rise to the notochord. The notochord in vertebrates is a transient embryonic structure. It plays an important role in the induction of the neural tube and in the differentiation

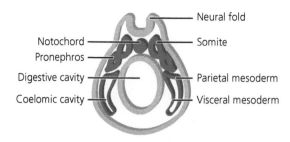

Figure 31.2 Cross section through a late neurula-stage frog embryo: Blue—ectoderm, red—mesoderm, yellow—endoderm.

of nearby tissues. Lateral to the notochord, the tissue known as **paraxial mesoderm** gives rise to a series of spherical cell clusters known as **somites** (Figure 31.2). Mesoderm that differentiates on the lateral sides of the embryo forms paired structures known as the **lateral plates**. Between the lateral plates and the axial and paraxial mesoderm, a long and narrow strip of tissue gives rise to the **intermediate mesoderm**.

The somites are the basis for almost all segmented structures in the body (with the exception of those formed by the neural crest and parts of the intermediate mesoderm). The process of somite formation or **somitogenesis** is the embryonic process of segmentation in vertebrates. Segmental organization in vertebrates is very different from what we've seen in annelids and arthropods. It is entirely internal, with no externally visible segmental structures. It is also mostly mesodermal, whereas segmentation in arthropods and annelids involves the ectoderm as well.

Shortly after somites form, they differentiate into a number of regions, each giving rise to a different organ system (Figure 31.3). The **sclerotome** will give rise to the axial skeleton or the vertebrae. The cells of the sclerotome will eventually encircle the neural tube, forming a protective bony housing for the central nervous system. In many vertebrates (but not in frog embryogenesis) the sclerotome also contributes to the dorsal portion of the ribs—lateral bony extensions from the vertebrae that protect the chest and the soft tissues within it. The vertebral column is composed of the sclerotome of pairs of somites, which enclose the neural tube from both sides. At the base of each vertebra is the remains

of the notochord, which degenerates throughout development. The vertebrae themselves are intersegmental, meaning they are formed by the anterior of one somite and the posterior of the somite in front of it.

The **dermatome** will give rise to the dermis, the mesodermal portion of the integument. While the dermis starts out segmental, it loses its segmental character, the contributions from the different somites fuse, and mature dermis is not segmented. Only the dermis of the back is formed from the dermatome, with the ventral dermis being formed by the lateral plates (see below).

The **myotome** will give rise to the segmental muscles of the body, while the **syndetome** will give rise to tendons. Trunk muscles in vertebrates, as well as the tendons that connect them to the bones, are segmental, although in many vertebrates the segmental organization is obscured by the presence of appendicular muscles and by the complex shape of the trunk. The limb musculature develops from the lateral portion of the somites in the region where the limbs emerge. The segmental organization of vertebrate muscles can be seen very clearly in fish, where most of the body is composed of repeated rhombus shaped muscles (Figure 31.4). Each of these individual muscles is derived from the myotome portion of one somite.

Formation of the coelom

The lateral plates are paired but non-segmental structures. They split into two components, with a space between them (Figure 31.2). The medial component of the lateral plates is known as the **visceral mesoderm**. It will surround the endoderm and fuse with it to give rise to all the mesodermal components of the digestive system (Figure 31.4). These include the smooth muscles, mesodermal cells within digestive glands, and other structures. The more lateral component of the lateral plates is known as the **parietal mesoderm**. It will give rise to all the remaining mesodermal structures not associated with the digestive system or the somites. This includes the mesoderm of the body wall including the ventral dermis, non-segmental muscles, ventral ribs, and various internal membranes. The parietal mesoderm is also the embryonic source of the circulatory system, including the heart, most blood vessels, and the cells of the circulatory system. Finally, the body wall mesoderm, which is derived from the parietal mesoderm, is the source of the skeleton and connective tissue of the appendages.

The space between the two components of the lateral plates will form the coelom. Recall that true coeloms are surrounded by mesoderm, and form between the mesoderm of the body wall and the mesoderm of the digestive system (see Figure 16.3). This is exactly the case in the vertebrate coelom. The coelom in vertebrates is unsegmented and has no somitic component (unlike the case in annelids where the coelom is made up of segmental cavities). It is divided into two separate cavities, the **abdominal** (or peritoneal) **cavity** and the **pericardial cavity**. The former holds most of the digestive system and the latter surrounds the heart and the lungs. Between the two cavities is a membrane known as the **septum transversum**. In amniotes, this membrane becomes muscular and turns into the **diaphragm**.

The visceral mesoderm remains connected to the rest of the body via a membrane formed by a fusion of portions of both sections of the lateral plate. This membrane is known as the **mesenterion** (Figure 31.3). The blood supply to the digestive system is via blood vessels that run along the mesenterion (Figure 31.4).

The digestive tract

The archenteron, or primitive gut, differentiates along the anterior–posterior axis to form different regions. The endodermal component differentiates to give various types of epithelia along different

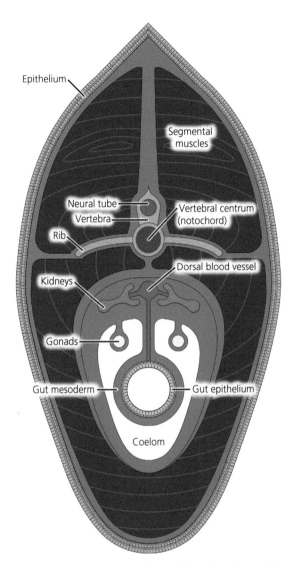

Figure 31.4 Cross section through an adult vertebrate, illustrated by a generalized fish.

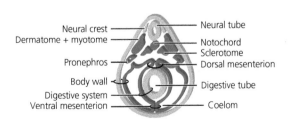

Figure 31.3 Cross section through a tailbud-stage frog embryo: Blue—ectoderm, red—mesoderm, yellow—endoderm.

regions of the gut. The mesodermal component (from the visceral mesoderm) differentiates to give thicker or thinner muscle layers in the more muscular and less muscular sections of the digestive system. The large digestive glands—the liver and pancreas—start out as out-pocketing from the endoderm of the digestive tube. Later in their development they incorporate elements from the visceral mesoderm to give compound structures with numerous cell types. The lungs also start as dorsal out-pocketing of the digestive system (note that they are not formed during embryogenesis in frogs). They then combine with blood vessels from the parietal mesoderm to give the adult structure that allows oxygenation of the blood.

The excretory system and gonads

We briefly discussed the structure of vertebrate kidneys in Chapter 28. The functional kidney of the frog tadpole is a very simple version, known as a **pronephros**. It is an anterior structure derived from the intermediate mesoderm. The pronephros is segmental, roughly corresponding to the anterior somites, with each segment including a single filtering unit or nephron. The pronephros appears as a transient embryonic structure in the embryonic development of most vertebrates. In fish and adult amphibians, it is replaced by a structure known as a **mesonephros**. The mesonephros is also formed from the intermediate mesoderm. It is more complex in structure and includes several nephrons bundled together. It is positioned posteriorly to the embryonic pronephros. Amniotes have a third type of kidney, known as the **metanephros**, although they have a transient pronephros and a functional mesonephros during embryonic development. The metanephros is a posterior structure, shaped like a bean. It includes many nephrons (up to a million in large animas) bundled together in a complex branched structure, with no evidence of segmental organization.

Vertebrate gonads are mesoderm-derived structures. They include both the support structures and nutritive elements of the germ cells, and the germ cells themselves. The germ cells are derived from a distinct embryonic tissue known as germ plasm. The gonads and kidneys occupy similar positions, dorsal to the coelom, but separated from the coelom by the parietal membrane. The vessels of the excretory system and of the reproductive system are linked embryonically and shared in some cases, but we will not go into the complex details of this relationship here.

The development of sense organs

The brain develops as an anterior expansion of the hollow nerve cord that occurs shortly after the end of neurulation. The expanded region differentiates into three regions as detailed in Chapter 29, giving rise to four brain ventricles: paired ventricles in the telencephalon, a third ventricle in the diencephalon, and a fourth ventricle in the medulla.

Many of the sense organs of the head and body start as thickenings of the ectoderm—placodes—that interact with the mesoderm to differentiate into mature sense organs. We will give a few examples to demonstrate. In eye development, the placode forms the lens of the eye. The eye itself is an extension of the brain that forms a cup-like structure with the lens in its center. The ear placodes form spherical structures that divide in two and differentiate to give the vestibular region of the ear (the semicircular canals) and the auditory region (the cochlea). The lateral line system starts as a series of placodes along the lateral body wall. The placodes sink in and merge to give the mature line system.

The development of the heart and circulatory system

The circulatory system starts out as a series of cell clusters known as blood islands. These fuse to give a series of tubular structures, with the largest one being the mid-ventral tube that will ultimately give rise to the heart and the medial blood vessels. The heart starts out as a muscular tube and begins beating almost as soon as it is formed, long before there is any actual circulation of blood. The heart tube bends and loops to the left, giving the characteristic asymmetry of the heart.

Blood cells also start out in the blood islands. They undergo a series of differentiation events, via a number of intermediate cell types, to give rise to the range of blood cells found in adult vertebrates. This process is known as **hematopoiesis** (or hemopoiesis) and is one of the best-studied processes of cell differentiation.

CHAPTER 32

Organismic biology in the twenty-first century

A variety of animal body plans

Throughout this book, we've reviewed a range of different animal groups and an assortment of components that make up their body plans. We've seen that animals have very different body plans. Some animals have relatively simple organization whereas others are fairly complex. We've also seen that there are differences in the diversity of different groups of animals. Some phyla are highly diverse with many thousands of described species, whereas others are small with only a few hundred species. Within this diversity, some taxa are conservative, with all members of the taxon being minor variations on a typical body plan, whereas others vary significantly and a typical body plan is difficult to identify. The reason for these differences in complexity, diversity, and conservatism in different animal taxa remains one of the great mysteries of biology.

This book highlights animal diversity while also looking for conserved organizational principles. Conserved aspects among different animals can be the result of shared ancestry, but they can also be due to similar fundamental requirements underlying the way an animal is structured and the way it functions. Disentangling the two sources of similarity can be difficult, and in this book, we've only scratched the surface of trying to identify which similarities are due to which underlying reasons.

Organismic biology as a discipline has waxed and waned in popularity and relative importance over the years. For much of the end of the twentieth century and even the beginning of the twenty-first century it suffered a decline and took a back seat relative to the more reductionist and mechanism-focused fields of molecular biology and genetics. Recently, organismic biology has regained its standing as a central component of modern biology, largely through the integration of mechanistic approaches into a broad organismic view. This modern take on organismic biology is now trying to tackle the questions of animal form and function outlined above. Of the different approaches that bring mechanisms into organismic biology there is one approach—evolutionary developmental biology, or Evo-Devo—that (in the opinion of the author) stands out among others. A brief introduction to this approach will be at the center of our final chapter.

Developmental biology as a unifying discipline

In Chapter 25, we saw that development is the process that specifies and generates the animal body plan. The developmental process "reads" the genome and translates it into morphology, as the fertilized zygote goes through the developmental process. This is, of course, not a straightforward or direct reading of the genome. There is no direct encoding of specific morphologies anywhere in DNA sequence that comprises the genome. Rather, the genome contains a series of instructions for activating the different components that constitute the developmental processes that generates morphology.

The evolution of morphology is thus the evolution of the process that generates the morphology,

Organismic Animal Biology. Ariel D. Chipman, Oxford University Press. © Ariel D. Chipman (2024). DOI: 10.1093/oso/9780192893581.003.0032

or in other words the evolution of development. Since every aspect of an animal's morphology is the result of developmental processes, we can trace the development of individual structures or systems, within a phylogenetic framework, as a basis for understanding how the process varies and how the resulting structure changes during the evolutionary process. This is in fact the approach we adopted for understanding the evolution of the vertebrate body plan in Chapter 31. It is important to note that this realization is not new. Most of what we presented in Chapter 31 has been known for over 100 years, and comparative vertebrate morphologists have been looking at development as a tool for understanding morphological diversity within vertebrates for at least that long. What then are the novel aspects of modern Evo-Devo?

A conserved developmental toolkit

By the late 1970s and early 1980s, developmental biology had adopted many of the tools of experimental genetics, leading to the birth of developmental genetics—the study of the genes involved in the developmental process. This was initially done on a very small number of organisms, most notably the fruit fly *Drosophila melanogaster* (see Box 32.1). During the early 1990s, genetic approaches were applied to the development of additional organisms, leading to a series of surprising discoveries. Despite the diversity of morphology in different animals, it emerged that there is a limited number of genes underlying development in all animals. The first conserved developmental genes to be discovered were the Hox genes, a series of genes involved in patterning the anterior–posterior axis in all bilaterian animals. It gradually emerged that Hox genes are not unique and that there are many such developmental genes that are conserved across a wide range of animals. These genes have been dubbed developmental toolkit genes, with the underlying metaphor implying that just like you can build a house or a cart with a limited set of tools (hammer, screwdriver, wrench, etc.), development can construct the entire diversity of animals using a limited number of genes.

Developmental toolkit genes mostly belong to two classes of genes: genes encoding transcription factors and genes encoding signaling molecules or components of signaling pathways. Transcription factors are proteins that bind to control regions of the DNA and activate other genes. Signaling molecules transmit information between cells, leading to the activation of genes in other cells. Transcription factors and signaling molecules work together to form complex networks of interactions among genes.

We can now take our earlier statement about development "reading" the genome to create morphology, and rephrase it using more precise wording. We can say that the developmental process is the sequential activation of networks made up of genes that encode transcription factors and signaling molecules, leading to a series of cell divisions, cell movements, and differentiation events that generate morphology.

Evo-Devo takes this mechanistic understanding and uses it to follow which aspects of development are conserved and which are variable in order to reconstruct the changes in developmental genes that underlie changes in morphology over evolutionary time. By identifying these genes and the developmental context in which they operate, evolutionary developmental biologists can reconstruct when and where in evolution specific aspects of the body plan first appeared and how they have been modified in different lineages.

A few examples that are relevant to topics we discussed in this book will serve to demonstrate these ideas. Genes involved in axial patterning, both anterior–posterior and dorso-ventral were discovered in the early days of Evo-Devo and were shown to be conserved among all Bilateria. Studying these genes in non-bilaterian animals can shed light on how axes first evolved in the early history of bilaterians. The Hox genes mentioned above are responsible for defining different regions along the anterior–posterior axis. Work on both arthropods and vertebrates showed how different taxa within these phyla use Hox genes to differentiate different regions of the body. Genes involved in the segmentation process of arthropods demonstrated that the segmented body plan is an ancient and conserved process within arthropods, but that despite some similarities, probably evolved independently in annelids.

Some of the most recent work in evolutionary developmental biology has tackled the origin of cell types. Scientists have identified cell-type specific genes or groups of genes that encode conserved components of the cell. For example, muscle cells have a set of conserved genes that are involved in the synthesis and organization of the muscle-specific proteins actin and myosin. Nerve cells have conserved specific genes involved in synthesizing neurotransmitters and in encoding the receptor proteins that neurotransmitters bind to. These approaches allow tracing the evolution and origin of cell types and allow us to answer questions about the evolutionary origin of specific tissues and organs and about their homology in different taxa.

Conserved networks and homology

During the early history of genetics, the view on gene function was that one gene was responsible for one character. Indeed, the first genes that were characterized fit this idea. There were genes for flower color, for seed shape, for the presence or absence of wings, etc. By the late twentieth century, it was clear that this view was overly simplistic, and the prevailing idea was that every gene had one function, but this function could be employed over and over in different structures and different contexts. Nonetheless, the view that linked specific genes with specific structures or characters remained prevalent among evolutionary biologists, and the involvement of related genes in similar structures in different organisms was seen as evidence for homology of these structures.

For example, similar genes were found to be active in the early development of the eye in distantly related animals, and this was seen as evidence for a common origin of eyes in all bilaterians. The presumed homology of all bilaterian eyes came as a surprise to most evolutionary biologists, since it was widely accepted, based on morphological arguments, that eyes evolved independently and convergently in different lineages. The genes found in common in the eyes of such diverse animals suggested that there may be some sort of very ancient link between all these eyes. Perhaps a common

ancestor with a simple eye composed of a single pigment cell gave rise to all the eyes we see today. Such an ancient link between morphologically dissimilar structures was called "deep homology."

This view of genetic similarity underlying homology is true to a certain extent but must be used with caution. There isn't a gene for every structure or character. There is no gene for the wing of a fly, for the parapodium of an annelid, or for the flame cell of a platyhelminth. All genes have numerous roles and all structures are the result of the activity of numerous genes. A more nuanced point of view, which is starting to take hold among evolutionary biologists, but has yet to be demonstrated in more than a handful of specific cases, sees gene regulatory networks (GRNs) as the conserved element. According to this view, homologous structures are patterned by a series of genes interacting in a conserved network. In other words, it is not the specific genes that are at the basis of a character or structure, but the conserved interactions between them.

The future of organismic biology

We are now at a crucial point in the history of organismic biology. The introduction of mechanistic thinking into organismic biology was spearheaded by the Evo-Devo revolution of the turn of the century but is now prevalent in most aspects of organismic biology. The genomics revolution of the second decade of the twenty-first century allowed the sequencing of the genome of numerous species, not only standard model organisms and species of economic importance. This came at the same time as an increased interest in studying members of under-studied and neglected taxa.

Organismic biology is shaking off its outdated image of a field studied in dusty museum basements. The leading universities in the world have active research programs in organismic biology, and a new generation of enthusiastic, young scientists are using cutting edge techniques to study age-old questions on the evolution and relationships of animals and on the structure and function of organs and organ systems.

This organismic renaissance has not gone unnoticed by the biological world as a whole, and there is an increased recognition of the importance of understanding the whole organism as a basis for any research program in biology. We hope that this book has given you the reader such a basis.

Box 32.1 The "big five" model organisms

Most biological research focuses not on diversity but on general principles. Studies are therefore mostly carried out on a small number of species that can be easily maintained in labs worldwide, allowing standardization of experimental procedures among researchers. These commonly used species are known as "model organisms." There are many species that are used in research labs, and their number is growing constantly. However, there are five species, often known as the "big five" that are used far more than any others and have been mainstays of biological research for many decades. These species have been chosen for practical reasons—size, ease of maintenance, and short generation time. They are often seen as representatives of their taxa, but this representation is problematic, since it is often the characteristics that make them suitable for the lab that also make them unusual within their representative taxa. The five main model species are listed below, with their experimental advantages and the main fields in which they are used.

Mus musculus—the house mouse. This is by far the most widely used model organism. It is used for testing whole-organism processes, usually in reference to human biology and disease, since it is a mammal, belonging to the same class as humans. Mice can be kept in small enclosures and have a short generation time for a mammal, and thus can provide rapid experimental results in genetic experiments. Mice are also used in studies in psychology and behavior due to their relatively complex social structure. Many genetic and molecular tools are available for manipulating mice experimentally.

Drosophila melanogaster—the fruit fly. *Drosophila* came to prominence in the first days of genetics in the early twentieth century, when a number of labs started rearing fruit flies in order to identify and study genetic mutations and gene activity. From the late twentieth

century, they have also been used extensively for studies in developmental biology. The first genes involved in developmental processes were discovered in *Drosophila*. Today, *Drosophila* is used for a very wide range of research in genetics, development, cell biology, and other fields. Numerous strains exhibiting specific mutations are available, and the ability to carry out gene manipulation experiments in *Drosophila* is unparalleled by any other species.

Caenorhabditis elegans—the nematode, usually given in the short form as *C. elegans*. The fixed number of cells in this species, as well as the fully characterized stereotypical development, makes this the most standardized model organism. *C. elegans* has been used in labs since the 1960s. It is used in almost all fields of biology from aging research to neurobiology, developmental biology, gene regulation and others. Its miniscule size means it can be reared in very high numbers and its short life cycle makes it useful for multi-generational studies.

Xenopus laevis—the African clawed frog. *Xenopus* first entered research labs in the 1930s as a pregnancy test. Injection of urine from a pregnant woman into a female frog leads to the release of the frog's eggs. The ease of obtaining eggs from this frog (synthetic hormones can be used instead of female urine) then made *Xenopus* a prime model for studies in developmental biology. Frog eggs can be obtained in larger numbers and more easily than almost any other vertebrate, and they thus became the main model species for vertebrate development (see Chapter 31). The eggs can also be used as expression vehicles for vertebrate genes and *Xenopus* is used for broader studies in molecular biology.

Danio rerio—the zebrafish. The most recent addition of this list to the model organism toolkit. Zebrafish have been used for studies in developmental biology and gene expression from the 1970s. They were among the first vertebrates to be used for genetic screens (much like *Drosophila* was more than half a century earlier). The scale of zebrafish use does not reach that of the other species mentioned, but it is the most experimentally tractable vertebrate species, making it important for many basic and applied questions that are relevant to human biology.

Index

Boxes are indicated by an italic *b* following the page/paragraph number.